我们在中国
——多样性的中国野生动物

WE LIVE HERE:
DIVERSE WILD ANIMALS IN CHINA

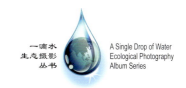
一滴水生态摄影丛书 / A Single Drop of Water Ecological Photography Album Series

陈建伟 著/摄
CHEN JIANWEI

中国林业出版社
China Forestry Publishing House

图书在版编目（CIP）数据

我们在中国：多样性的中国野生动物 / 陈建伟著、摄. -- 北京：中国林业出版社，2021.9
ISBN 978-7-5219-1176-3

Ⅰ. ①我… Ⅱ. ①陈… Ⅲ. ①野生动物－艺术摄影－中国－现代－摄影集 Ⅳ. ①J429.5

中国版本图书馆CIP数据核字(2021)第097226号

审图号：GS（2020）384号

中国林业出版社·自然保护分社（国家公园分社）
策划编辑：肖静
责任编辑：肖静　何游云
特约编辑：田红
英文翻译：柴晚锁　陈肖潇　张宇彤　张雪辰　马远　芒孜热木·克日木
英文审校：郭瑜富
图片编辑：崔林
装帧设计：崔林

出版发行：中国林业出版社（100009　北京市西城区刘海胡同7号）
　　　　　http://lycb.forestry.gov.cn
电话：（010）83143577
E-mail：forestryxj@126.com
印刷：北京雅昌艺术印刷有限公司
版次：2021年9月第1版
印次：2021年9月第1次
开本：787mm × 1092mm　1/12
印张：19
字数：300千字
定价：320.00元

未经许可，不得以任何方式复制或抄袭本书之部分或全部内容。
版权所有　侵权必究

谨把此书：献给所有为野生动物保护作出努力的人们！

This book is dedicated to all those who have been working hard for the conservation of wildlife!

屹立雄视·陕西佛坪
A serious look from above — Foping, Shaanxi

大熊猫不仅是中国自然保护的旗舰物种和中国国宝，更是世界自然保护及和平友好的使者，全球最大的环保组织之一——世界自然基金会（WWF）的知名标志来自于此。
The giant panda (*Ailuropoda melanoleuca*) is not only a flagship species and national treasure of China, but also an ambassador of natural protection as well as world peace and friendship. As one of the world's largest environmental protection organizations, the well-known logo of the World Wide Fund for Nature (WWF) comes from it.

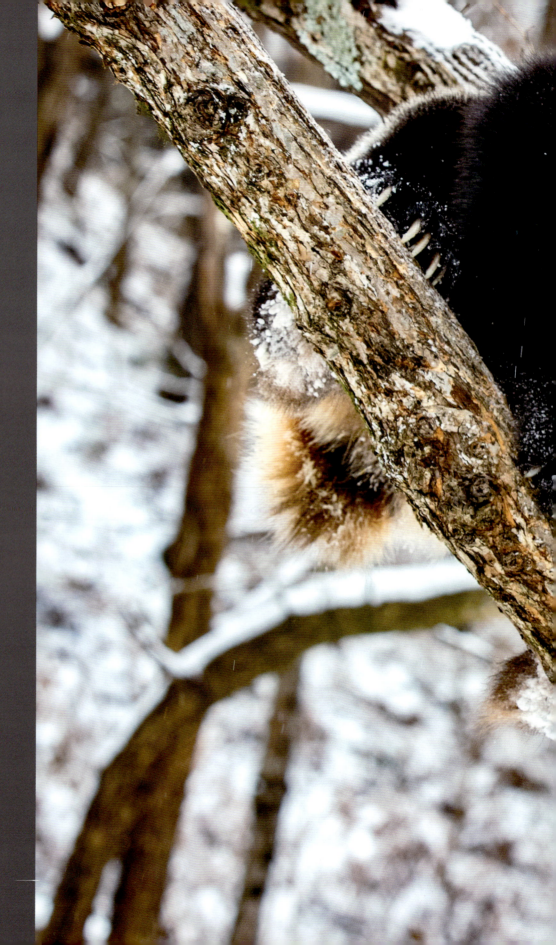

晨中仙鹤·黑龙江扎龙
Fairy crane at dawn — Zhalong, Heilongjiang

丹顶鹤是中国文化中的"仙鹤"，是美丽、飘逸、长寿、吉祥和高贵的象征。
Red-crowned crane (*Grus japonensis*) is called "fairy crane" in Chinese culture. It is a symbol of beauty, elegance, longevity, auspiciousness and nobility.

自然精灵·湖北神农架
The elves in nature — Shennongjia, Hubei

华丽灵慧的川金丝猴家族中洋溢着的是满满的和谐美好气氛。
Harmonious and happy atmosphere prevails this gorgeous and intelligent Sichuan golden monkey (*Rhinopithecus roxellana*) family.

← 喜迎日出·山东长岛
Celebrating the sunrise — Changdao County, Shandong

西太平洋斑海豹是唯一能在中国黄海和渤海繁殖的鳍足类海洋水生动物，是黄渤海海洋生态系统的关键物种。沐浴在温暖阳光下的它显得格外幸福满足。

The spotted seal (*Phoca largha*), the only pinniped marine aquatic animal that can breed in the Yellow Sea and the Bohai Sea in China, is the key species of the marine ecosystem of the Yellow Sea and the Bohai Sea. It is extremely happy and content in the warm sunshine.

序一

很高兴为这本献给联合国《生物多样性公约》第十五次缔约方大会的书作序。这次公约大会我担任中方首席专家。我是从事保护生物学研究的，保护生物学就是研究生物多样性保护的新兴学科，是一门基础科学与应用科学、自然科学与社会科学双交叉的综合学科。陈建伟先生和我很早就认识，他是技术出身的行政管理人员，有着坚实的基层业务基础和丰富的野外调查监测经历，还曾担任过全国第一次野生动物资源调查的技术总负责人，后来又到国家最高层面的野生动物行政主管部门担任领导职务，具有自然科学和社会科学相结合的经历，我们常在一起讨论保护生物学方面的很多问题。

如科学家所描述的，我们人类正经历地球历史上第六次物种大灭绝时代，开启者是人类自己，因此能改变的也只能是我们自己。

这本书采用了生态摄影的新理念，用精彩的图片和洗练的文字，给我们讲述了一个又一个生动的中国野生动物故事，不仅宣传了中国野生动物保护的成绩，提出了存在的问题，还进行了深度的思考，是一本非常难得的对于大众学习、认识中国野生动物保护的优秀科普书。此书在公约大会上进行发布，也将让各国代表对于中国野生动物的保护有一个更深、更全面的了解，一定能够在国内、国外产生很好的生态效益和社会效益。

我认为，这本书有几个创新点值得特别提出。

一、该书采用的动物地理区划并没有拘泥于现成的通用的中国动物地理区划，而是在此基础上结合两次全国野生动物资源调查数据细化而形成的成果应用，这更加符合当前中国实际，是有坚实科学数据支撑的。

二、介绍中国野生动物的书不少，但用生态摄影的新理念、新方式来全面讲述中国野生动物故事的还是首次。生态摄影强调的是思想性、科学性和艺术性的统一结合，做好并不简单。中国有如此丰富多样的动物，要在一本仅有200多页的生态摄影集中讲好中国的野生动物故事，既要面上兼顾，又要重点突出；既要展示野生动物美丽可爱的一面，又要赋予生态思考的内涵；既要充分肯定保护工作的成绩，又不能回避存在的问题。这都挺考验作者功力。

三、本书打破了将野生动物分为陆生、水生的现行概念，而是从生态系统、保护生物学的角度出发来介绍，减少了人为的划分对于生命共同体理解的影响。为了让读者更好地了解野生动物的知识，书的后面还专门将物种的中文名、拉丁学名以及《国家重点保护野生动物名录》《CITES附录》和《IUCN红色名录》级别一一列出，便于读者查询和理解。

四、书中讲述的大大小小的野生动物故事，始终融合贯穿了保护生物学的理念，如旗舰种、关键种、伞护种、指示种以及物种与物种的关系、物种与生态系统的关系、生态平衡、生态危机、可持续发展等，也希望这些理念成为公众关注的焦点。这方面这本书是竭力去做了。可以这么说，仅凭一人之力来编辑出版这本书是有相当难度的，懂业务不懂摄影的出不来，懂摄影不懂业务的也出不来，既懂业务又懂摄影但不了解保护生物学的还是出不来。

"生态文明：共建地球生命共同体"是本次公约大会的主题。对于人与自然到底应该是个什么关系的终极问题，联合国认可的行动领域是"基于自然的解决方案"，我们中国秉承的是"天人合一"和"师法自然"的哲学理念。而新型冠状病毒肺炎在全球的肆虐也给野生动物保护赋予了更加深刻的内涵。疫情警示人类，病毒没有国界，疫情不分种族，在病毒面前，一国抗疫成功不算成功，只有各国都成功了才是真正的成功。全人类就是一个命运共同体！整个地球生物圈也是一个不可分割的命运共同体！

在此，我也希望借这个平台，真切地祝贺即将召开的联合国《生物多样性公约》第十五次缔约方大会取得预想的成功！我想，作者和我也是同一个心愿。

中国科学院院士 魏辅文

2021年6月22日

Foreword I

I am very delighted to write this preface for a book that is to be presented as a gift to the upcoming fifteenth meeting of the Conference of the Parties to the *Convention on Biological Diversity* (UN CBD COP15), for which I will serve as the chief technical advisor of the Chinese delegation. My major field is conservational biology, an emerging interdisciplinary subject that focuses primarily on the conservation of biodiversity and one that requires simultaneous use of approaches typically adopted in both fundamental and applied sciences, and in both natural and social sciences. I have long known Mr. Chen as a highly professional and competent executive official who also has a very strong background in the technical arena. In addition to his expertise in working on grass-root posts and his rich experiences in field investigations and wildlife survey, he also worked as the chief technical official during the first nation-wide inventory on China's wildlife resources before he assumed his administrative responsibilities in the country's highest decision-making organization concerning wildlife management. He combines the knowledge he has obtained in both natural and social sciences into an integrated whole that has enabled him to make quite a lot of impressive achievements in his career. We have frequently been consulting each other on some key issues in the field of conservational biology.

Just as some scientists have described, we human beings are currently undergoing the sixth mass extinction of species that the Earth has been subjected to in its history. Since it is us, the so-called *Homo sapiens*, who have started this alarming trend of mass extinction, we naturally have the moral obligation to take actions to alter its course.

The new concept of eco-photographing, which makes use of both vivid pictures and brief but insightful language, is adopted throughout the book to tell the readers some touching stories related to wildlife species in China. These stories not only celebrate on the impressive achievements that China has made in its wildlife conservation efforts, but also aim to identify the problems that still exist in this aspect through the author's insightful reflections on our work. This book will help the general public to know about and hence gain a better understanding about the current state of wildlife conservation in China, and therefore constitutes a high commendable reading in popular sciences. Besides, its release, which coincides with the convention of the UN CBD COP15, will also play a constructive role in giving foreign delegates to the conference a better and more informed understanding about the efforts as well as achievements that China has made in wildlife conservation, which will in turn generate desirable ecological and social impacts in both China and the international community.

I deem this book particularly valuable and credit-worthy for some innovative observations that it raises.

Firstly, instead of being limited to the confinement of existing system that has been conventionally adopted in the division of zoogeographical regions in China, the system proposed in this book takes into consideration the latest data generated from the two nation-wide inventories on wildlife resources in the country and therefore is more detailed, accurate and solidly-based on scientific data, hence presenting a more realistic picture about the current state of wildlife conservation in China.

Secondly, books featuring wildlife in China are by no means scarce, but this book marks the first attempt in adopting such an innovative and ground-breaking approach as eco-photographing for this purpose. Eco-photographing calls for perfect blending of impregnable thoughts, rigorous science and aesthetical art and therefore is indeed a challenging task. With the rich wildlife diversity of China and the limited space in an eco-album of more than 200 pages, how can we ensure that a good job be done in presenting the readers with stories that are at the same time all-covering and well-focused in species selection, conducive to highlighting the cute and interesting aspects of wildlife and thought-provoking in terms of ecological insights, celebrating on the achievements without being evasive to problems that still exist? All these considerations make up tremendous challenges to the author's competence and professional finesse.

Thirdly, the author of this book breaks deliberately away from the conventional approach that divides wildlife into terrestrial and aquatic species. Instead, he organizes his stories from an angle that is solidly based on ecological system and conservational biology so as to reduce to the greatest possible extent the negative impacts that the existing arbitrary delineation between wildlife species might have on our understanding about the notion that all life on earth is of a community of shared life. For easier access to and better understanding about knowledge in this field by the readers, the author has included as an appendix to the book a list of the species mentioned in the album, including their names in both Chinese and Latin, as well as their respective status in the *List of Key Protected Wild Animals of National Significance*, the *CITES Appendices* and the *IUCN Red List of Endangered Species*.

Fourthly, the stories included in this book are aimed to illustrate the key concepts in conservational biology — indicator species, key species, umbrella species, flagship species and *etc.*, the relationship between different species as well as that between individual species and the whole ecosystem, ecological balance, ecological crisis, sustainable development — and hence enhance people's awareness about them. The author has indeed done a very good job in this regard. Frankly speaking, it is not an easy task for anyone to have come up with such a book single-handedly, for it is simply not possible to accomplish this if a person has strong background in only wildlife management or photographing but not in the other aspect, or excels in both but knows little about conservational biology.

The theme of the UN CBD COP15 is "Ecological Civilization — Building a Shared Future for All Life on the Earth". In response to the utimate question concerning the ideal human-nature relationship, the UN endorses nature-based solutions, which is in essence much similar to the philosophical approaches that we Chinese typically adhere to — "harmonious integration between human beings and nature" and "drawing on the laws of the natural world". The COVID-19 pandemic that plagues the world has further highlighted the importance as well as the urgent needs for protecting wildlife by drawing our attentions to the alarming reality that virus has no respect for neither national borders nor the boundaries between races, that the success of any individual country in containing the disease will count little if other countries fail in the task. The whole humankind makes up a community of shared future! The entire biosphere of this planet makes up an indivisible community of shared future!

I would also like to avail myself to this platform to extend my most heart-felt wish that the upcoming UN CBD COP15 will turn out to be a great success. And I believe this wish is also shared by the author of this book.

WEI Fuwen, CAS Academician

June 22, 2021

序二

众所周知，森林、湿地、荒漠、草原、海洋等组成了地球上最主要的自然生态系统，而野生动植物是各类自然生态系统最重要的组成部分，也是人类的同行者和不可分离的朋友。这些生态系统和野生动植物共同组成的生物多样性是人类社会赖以生存和发展的最重要的基础。

中国政府高度重视生物多样性保护工作，在自然保护领域开展了一系列卓有成效的工作。建立了以国家公园为主体的自然保护地体系，国家公园、自然保护区和各类自然公园等达到1.18万处，有效地保护了自然生态系统的原真性、完整性。截至2021年4月，我国自然保护地总面积占国土陆域面积的18%，占管辖海域面积的4.1%，有效保护了我国90%的陆地生态系统类型、85%的野生动物种群、65%的高等植物群落，在我国保护生物多样性、改善生态环境质量和维护国家生态安全方面发挥了重要作用。

我一生都在从事自然保护工作，为了彰显中国自然保护工作的成就，以生态摄影的创新形式，先后著、摄并出版了：

《一滴水生态摄影集》（2008年）——作为中国政府"2008北京奥运会"送给各国贵宾的指定礼品，获组织委员会授权使用"2008北京奥运会"的会徽及五个福娃形象，并作为奥运文化遗产收藏；

《多样性的中国森林》（2011年）——作为中国政府送给"2011国际森林年"联合国大会的礼品，获"梁希科普奖"；

《多样性的中国湿地》（2014年）——为"中国第三届湿地文化节"礼品；

《多样性的中国荒漠》（2017年）——作为中国政府给"联合国《防治荒漠化公约》第十三次缔约方大会"的礼品，赠送给186个国家和20多个国际组织的代表，获第五届中国科普作家协会优秀科普作品奖金奖和科学技术部推荐"2019年全国优秀科普作品"。

习近平主席最近指出："我们要同心协力，抓紧行动，在发展中保护，在保护中发展，共建万物和谐的美丽家园"，并提出"山水林田湖草沙是生命共同体"的理念。为向世界各国介绍中国的野生动物状况，以及对野生动物保护工作的思考，继前面出版的四本书的基础上，我著并摄了以介绍中国野生动物为主题的《我们在中国——多样性的中国野生动物》，为首次在中国召开的联合国生物多样性第十五次缔约方大会献礼，让国际友人通过一幅幅精彩生动的图像故事和文字诠释了解中国的野生动物保护工作和生物多样性保护成效，为大力宣传"生态文明：共建地球生命共同体"的公约大会主题服务。

"基于自然的解决方案"是联合国认可的行动领域，与我国秉承的"天人合一"和"师法自然"的哲学理念琴瑟相和。在当前紧迫的新形势下，我们每一个人都要积极行动起来，用自己擅长的方式为保护野生动物作一分努力，为中国和全球生物多样性的保护作出应有的贡献！

陈建伟

2021年6月22日

Foreward II

It is widely acknowledged that forests, wetlands, deserts, grasslands and oceans jointly make up the majority of natural ecological system, in which wild fauna and flora species are the most important composing elements of various natural eco-systems as well as indispensable friends of human beings. Biodiversity, which includes all these eco-systems and wildlife thereof inhabit, is the most critical foundation on which human society relies for its survival and development.

The Chinese government has consistently placed biodiversity conservation in top priority and has carried out a series of work in nature conservation that have led to impressive achievements. With the development of a natural protected area system highlighting the central role that the national parks play, the total number of protected areas — including national parks, nature reserves and other nature-based parks — across the country currently stands over 11,800, playing an effective role in maintaining the integrity and intactness of all types of natural ecosystems. As of April 2021, the total size of various protected areas in China accounts for 18% of its land territory and 4.1% of the oceans that fall within its jurisdiction. In other words, over 90% of land ecosystems, 85% of wildlife populations, and 65% of higher plant communities of the country are now under sound protection, laying down a solid foundation for wildlife conservation, improvements in the quality of eco-environment, as well as for safeguarding national ecological security.

As an active preacher as well as a practitioner who is, throughout my life, actively engaged in the protection and conservation of nature, I have so far published, in the form of innovatively-edited eco-albums, several books that celebrate on the great achievements that have been made by China in its endeavor in this regard, which are listed as follows:

A Single Drop of Water Ecological Photography Album (2008), which was presented as a designated gift from the Chinese government to the distinguished foreign guests who took part in the 2008 Beijing Olympic Games, for which authorization was granted by the Organizing Committee of the Games for the official emblem and the images of Fu Wa that are mascots of the Games to appear on its cover, and which was included into the collection of Olympics heritages;

The *Diverse Forests of China* (2011), which was presented as a gift from the Chinese government to the UN General Assembly to celebrate "the International Year of Forests 2011", and which was awarded the Liangxi Award for Books of Popular Science;

The *Diverse Wetlands of China* (2014), which was presented as the official gift for the Third Festival of Wetlands of China;

The *Diverse Deserts of China* (2017), which was presented as a gift from the Chinese government to delegates from 186 countries and over 20 international organizations to COP13 of the UN Convention on Combating Desertification, and which also ranked among the Gold Medal winners of the Fifth Best Books Award in Popular Science presented by the China Science Writers Association, as well as among the 2019 List of Best Books in Popular Science recommended by the Ministry of Science and Technologies of China.

President Xi Jinping recently urged us "to take prompt and coordinated actions in responding to the demands of both nature conservation and social-economic growth so as to build our homelands into a beautiful place where all forms of lives can co-exist harmoniously and flourish". He also put forward and stressed on the idea that "views mountains, rivers, forests, farmlands, lakes, grasslands and deserts as an integrated community of shared life". To update the global communities about the present situation and latest approaches and achievements that China has made in wildlife protection and conservation, I am pleased to present this current book, *We Live Here: Diverse Wild Animals in China*, as a special gift to the UN CBD COP15 that is about to unveil for the first time in its history in China. It is sincerely hoped that it will help our foreign friends to gain, through the carefully selected pictures and the brief words that explain the stories behind these pictures, a better and more in-depth understanding about the remarkable achievements that China has made in wildlife protection and biodiversity conservation, hence playing an active role in raising people's awareness about the theme of this grand conference, that is, Ecological Civilization — Building a Shared Future for All Life on the Earth.

The UN-endorsed "nature-based solutions" resonate harmoniously with the ideas deeply embedded in the time-honored Chinese philosophy and traditional wisdom of the Chinese people, such as the belief that "humankinds and the universe co-exist harmoniously as an integrated whole", the proposal that "human beings need to draw on nature so as to better adapt themselves to the surrounding environment", and many others. Faced with the current pressing situation that threatens our very survival, each of us must respond resolutely and make his best efforts to contribute his due share in biodiversity conservation in both China and the rest of the world.

CHEN Jianwei

June 22, 2021

目录
Contents

导言 / Introduction **22**

东北区 / North-east Region **31**

蒙新区 / Inner Mongolia and Xinjiang Region **83**

华北区 / North China Region **59**

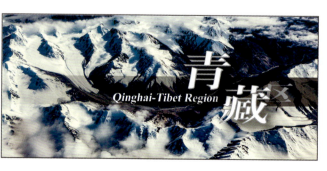
青藏区 / Qinghai-Tibet Region **109**

西南区 *South-west Region* **127**

华南区 *South China Region* **179**

华中区 *Central China Region* **155**

物种名录 / List of Species **216**

后记 / Epilogue **225**

导言
Introduction

为迎接2021年在中国昆明举行的"联合国《生物多样性公约》第十五次缔约方大会"，为了宣传中国异常丰富多样的野生动物，让广大国民更加清晰、全面地了解自己国家的野生动物概况，也为了向世界各国介绍中国的野生动物状况及保护成就，包括面临的问题和挑战，特推出这本以介绍中国野生动物为主题的生态摄影集：《我们在中国——多样性的中国野生动物》。

丰富多样的中国野生动物

中国的野生动物种类异常丰富，仅脊椎动物就有6500多种，在世界各国中名列前茅。其中，哺乳动物736种，位于世界第一；鸟类1445种，位于世界第五；两栖类450种；爬行类473种；鱼类2800多种（包括淡水和海洋）。特别是大熊猫、朱鹮、川金丝猴、滇金丝猴、黔金丝猴、扬子鳄、藏羚、普氏原羚等数十种珍稀濒危野生动物为中国所特有。

中国野生动物种类异常丰富多样的原因，在于中国有世界上最为多种多样的自然地理气候及自然生态系统。东亚季风气候影响和青藏高原的隆起，大大改变了中国的气候格局，使北纬30°上下往往成为荒漠的地带变成了温湿的亚热带，使中国这块土地上产生了世界上面积和跨度都较大的亚热带。青藏高原的东南部，也出现了纬度高达29°的热带雨林，超过热带23°26′纬度上限5°多，直线距离达600多千米。水平方向从高纬度向低纬度走，跨越寒温带、中温带、暖温带、北亚热带、中亚热带、南亚热带、边缘热带、中热带到赤道热带9个气候带。垂直方向从海平面往高海拔方向攀升，历经赤道热带、边缘热带、南亚热带、中亚热带、北亚热带、高原温带、高原亚温带、高原亚寒带和高原寒带9个气候带。复杂多样的自然地势，加之很多动物物种在地球的第四冰川期遗存了下来，因而，中国大地才有如此丰富多样的野生动物。放眼全球，比中国纬度高的大国，由于气候过于寒冷，其生物多样性远不如中国；和中国具有大致相同纬度的国家，由于没有巨大海拔差产生的复杂生态系统，也没有如此丰富的生物多样性；纬度低于中国但生物多样性丰富度能够超过中国的寥寥无几。我在前面出的四本书里，已经将中国森林、湿地、荒漠等各种不同生态系统的复杂多样介绍给了大家，异常丰富多样的生态系统必然带来异常丰富多样的野生动物物种，丰富多样的它们因此诞生、成长、繁衍在中国这块土地上。

动物地理区划

为了将如此丰富多样的中国野生动物介绍清楚，地理区划是前提。在世界动物地理区划中，中国跨越了古北界和东洋界，这两大动物界以淮河—秦岭和青藏高原东南缘连线划分，大概各占半壁江山。张荣祖先生在此基础上将中国划分为7个区（参见2011年出版的《中国动物地理》），其中，东北、华北、蒙新和青藏区4区属古北界，西南、华中和华南区3区属东洋界。由于中国的自然地理本底相当错综复杂，自然生态系统极其多样，野生动物种类繁多，受当时做区划的条件和技术手段的限制，加上数据支持不够甚至很多地方缺乏数据，7个区的边界划定还是比较粗放的。

随着中国整个社会经济的快速发展，全国野生动物保护和管理对于摸清资源家底的迫切性提升，1995—2003年，中国启动了第一次全国陆生野生动物资源调查，当时我就担任技术总负责人。之后的2011年，又启动了第二次全国陆生野生动物资源调查，这种全国统一规划、统一技术标准、统一时间的大规模调查，在世界大国中是没有先例的。两次全国陆生野生动物资源调查积累了大量的数据，摸清了很多野生动物分布的实际状况，这就为在原来野生动物区划方案上做进一步的区划调整奠定了坚实的基础。新修定的《全国陆生野生动物生态地理区划》将全国共划为7个区18个亚区54个动物地理省和239个地理单元。其中，较大的调整有：如柴达木盆地在蒙新区扩大了范围；又如，由于横断山区存在南北走向的高山峡谷，青藏区和西南区、西南区和华南区之间区划边界做了较细的调整，由粗放平滑线改成了多次来回的折返线；再如，原青藏区的藏东南部分区域划进了西南区，等等。因此，本书采用的全国野生动物7个区是在张荣祖先生区划的基础上，加上两次全国陆生野生动物资源调查的数据库支持形成的。

海洋动物的区划是个特例，目前，还没有任何中国的海洋动物区划。为了方便综合表述，本书只做了简单的处理，将七个大区中涉及海洋的东北、华北、华中、华南区之间的分界线往海洋延伸，自然形成了海洋动物的归属区。

关于陆生和水生野生动物

我国历来从管理上甚至包括科研监测体系上都把全国野生动物分为陆生野生动物和水生野生动物两大类。本书希望将这个界限打破，也是希望这个人为的划分不要影响到它们同处于一个生态区的客观事实。在这七个大区里，将生活在同样领域里的野生动物都一起展示，不再人为区分陆生或者水生野生动物。实际上，很多野生动物是分不开的，不仅是因为它们都生活在同一个生态系统中，相互依存、互为因果、不可分割，而且很多动物尤其是两栖爬行类动物实际上是不可能、也不应该将它们人为分为是陆生或是水生的。

编写宗旨及图片选择

作为《一滴水生态摄影丛书》的第五本书，本书继续秉承"用摄影语言讲好中国生态故事"的基本原则，努力争取做到一张照片讲一个动物的小故事，两三张照片讲一个动物的中故事，一组照片讲一类物种的大故事，一个章节讲一个动物区的整体故事，整本书讲好中国野生动物的总故事。

当然，这是竭尽能力去努力的方向，做的过程中差强人意的地方往往不少。编写过程中，也常常有几个问题困扰自己，譬如，生态摄影除了应用图片尽可能多地告诉读者信息以外，往往需要一段简洁的文字说明和画龙点睛的题目，如何在有限的篇幅内交代清楚物种的学名、濒危珍稀程度、生态关系、生态思考、拍摄地点等；又如，一个不漂亮也不起眼的物种也许是生态系统中的一个重要物种或者一个有生动故事的物种，是否采用；再如，每个物种的照片占比，说明文字的篇幅大小，图片量与文字量的比例调配，以及大故事和小故事的比例控制，等等。这些都是在编写过程中始终伴随着我的问题。

另外，和前面出版的四本介绍生态系统的画册不一样的是，介绍野生动物必须具体、细致、准确到每一个物种，而生态摄影集又不能做成野生动物图鉴。由于篇幅有限，一个物种往往只能采用一张照片来显示，每一张照片既要正确介绍一个物种还要讲生态故事，这就存在一个取舍问题，有时为了说明问题也许就要牺牲光影效果好的照片。总之，出一本中国的野生动物生态摄影集，且冠以多样性，要综合考虑的问题确实很多，这是完成这本书历时3年多的原因。当然，这并不包括过去几十年照片的拍摄和资料的积累。本书采用的照片是从浩瀚的几十万张照片中挑选出来的，且绝大多数都是野外拍摄的，只有极少数如老虎、熊、海洋生物的个别照片，在野外确实无法拍到，但为了说明问题，只好采用了在救护中心、野生动物园、海洋公园拍摄的照片，这种情况全书总共不超过10张，并且都标明了拍摄地点。

要在一本仅200多页的生态摄影集中讲好中国的野生动物故事，既要有思想性、科学性、艺术性还要兼顾多样性，目前并无先例可循。因此，反反复复挑来选去就是一个不可回避的必然过程。生态摄影追求的是集思想性、科学性和艺术性为一体的摄影艺术呈现，我既不能仅挑选光影效果好的照片，也不能只挑选有故事性而缺乏科学性的照片；既不能只追求艺术性而忽略思想性，也不能回避中国野生动物保护面临的挑战和存在的问题；既要有个人经历和生态思考的融入，也要尽量展示人与自然是生命共同体的内涵。我当然希望挑出的每一张照片都既有思想性、艺术性又有故事性，还要符合排版的整体设计效果，但事实告诉我，这是非常困难的。只有较好没有最好，综合权衡利弊常常伴随的就是徘徊犹豫、改来改去，使我深感自己能力水平之有限。

生态文明：共建地球生命共同体

这套《一滴水生态摄影丛书》，无论是已出版的以生态系统为主题的四本，还是以物种为主题的这本，贯彻始终的原则是致力于用图文并茂的生态摄影形式，将内容科学、生动、直观地呈现给大家，把人类和野生动物同是一个生命共同体的理念贯穿于本书编辑的整个过程。为了帮助大家更好地认识并保护好中国野生动物，在充分肯定成绩的基础上，本书并没有回避中国野生动物保护面临的挑战和存在的问题，同时也结合了自己的相关经历，提出了自己的思考，以便比较全面、客观地讲好中国的野生动物生态故事，为做好中国野生动物的科学普及工作，为提高全民族的生态文明水平尽到自己应尽的责任。

世界自然保护联盟（IUCN）在2019年7月公布了最新版的《濒危物种红色名录》，对全球濒危物种保护状况进行了全面评估。这次新增了7000个物种列入该名录，使得该名录收录的濒危物种首次超过10万个，其中，共有28338个濒危物种面临灭绝威胁。IUCN强调，人口不断增加、经济发展和全球变暖引起的自然环境急剧变化，加剧了脆弱动植物的生存竞争。其最新科学评估报告提出，目前，世界上约有1/4的动植物的生存受到威胁，约有上百万物种在几十年内面临灭绝！近来，更有科学家推断，根据当前地球物种的消失速度推断，地球生物可能要面临第六次生物大灭绝！

联合国《生物多样性公约》是一项由世界196个国家签署的具有法律约束力的公约，旨在保护濒临灭绝的动物和植物，最大程度地保护地球上多种多样的生物资源，以造福于当代和子孙后代。即将在中国昆明召开的"《生物多样性公约》第十五次缔约方大会"的主题是"生态文明：共建地球生命共同体"。这一大会主题，对引导国际社会保护生物多样性的政治意愿，推进全球生态文明建设，努力达成《生物多样性公约》提出的"到2050年实现生物多样性可持续利用和惠益分享，实现'人与自然和谐共生'的美好愿景"具有重要的战略意义。这次大会将审议《2020年后全球生物多样性框架》，敦促国际社会及各国采取必要的步骤和措施来扭转生物多样性的丧失，并确定2030年全球生物多样性新目标。

中国是世界上生物多样性最丰富的国家之一，中国提出"山水林田湖草沙"综合治理和绿色发展的理念、划定生态保护红线、建立已超过国土面积18%的各类自然保护地，使众多自然生态系统和大多数重要的野生动植物种群得到保护，并且将要为2030年全球生物多样性新目标去努力奋斗。应该这样说，"生态文明：共建地球生命共同体"这一主题顺应了世界绿色发展的潮流，表达了全世界人民共建、共享地球生命共同体的强烈愿望和心声，是中国人民愿意竭尽全力去实现的美好愿景，当然，也是我献上这本书的初衷。

It is a great pleasure for me to present *We Live Here: Diverse Wild Animals in China,* an ecological photography album featuring the fascinating wild animals to which China is home. Written and published as a gift to the 15th Conference of the Parties (COP 15) to the United Nations *Convention on Biological Diversity* (CBD) to be convened in Kunming in 2021, this book aims to provide the Chinese as well as the global community with a comprehensive understanding about the rich diversity of wild animal populations inhabiting China by presenting the readers with a clear and holistic picture of the status quo, the great achievements in China's endeavor for wildlife conservation, as well as the challenges and problems that the country is still faced with in this regard.

Rich Diversity of Wild Animals in China

China is home to an extraordinarily rich diversity of wild animal species and is among the top countries in the world in this regard, with the number of vertebrate species alone amounting up to over 6,500. Specifically, it has 736 mammal species, ranking the first worldwide; 1,445 bird species, ranking the fifth in the world; 450 amphibian species; 473 reptile species and over 2,800 fish species (including both freshwater and marine ones). Around a dozen of rare and endangered wild animals are endemic to China, such as the giant panda (*Ailuropoda melanoleuca*), crested ibis (*Nipponia nippon*), golden monkeys (*Rhinopithecus roxellana, Rhinopithecus bieti, Rhinopithecus brelichi*), Chinese alligator (*Alligator sinensis*), Tibetan antelope (*Pantholops hodgsonii*) and Przewalski's gazelle (*Procapra przewalskii*) to name a few.

The existence of the most variegated geographical climates and natural eco-systems in China explains the abundance and huge diversity of wild animals in the country. Due to the influence of the East Asian monsoon climate and the elevation of the Qinghai-Tibet Plateau, the climate patterns of China are significantly changed. As a result, the area situated close to 30° northern latitude in China, which would otherwise often be deserts in other parts of the world, is turned to a mildly humid subtropical zone and hence bestowed China with a subtropical region that is fairly large in terms of both its size and range even when judged from worldwide perspectives. Tropical rainforests are found at the south-eastern part of the Qinghai-Tibet Plateau that falls at 29° northern latitude, which is approximately 5°, or over 600 km in linear distance, higher than the normal boundary of the tropical climate zone, i.e., 23°26′ northern latitude. Horizontally,

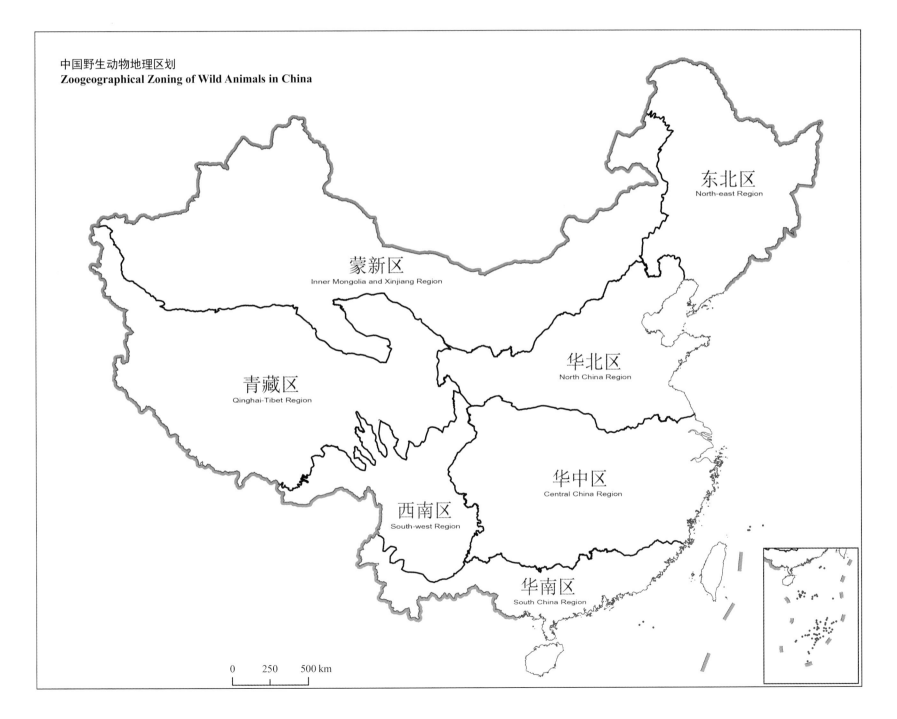

China covers nine climate belts that range successively from the high latitude to the low latitude, namely: the cold temperate zone, the mid-temperate zone, the warm temperate zone, the northern subtropical zone, the mid-subtropical zone, the southern subtropical zone, the marginal tropical zone, the mid-tropical zone, and the equatorial tropical zone. Vertically, the country covers nine climate belts that range successively from the sea level to high altitudes, namely: the equatorial tropical zone, the marginal tropical zone, the southern subtropical zone, the mid-subtropical zone, the northern subtropical zone, the plateau temperate zone, the plateau sub-temperate zone, the plateau subfrigid zone and the plateau frigid zone. Thanks to the highly variegated and complicated natural topography, as well as to the many species that had survived the fourth glacial period, the vast areas of China is able to become the home of such a huge diversity of wild animals. A quick look at the countries on the planet will show that, so far as biodiversity is concerned, those situated at latitudes higher than China typically fall far short because of too chilly climate, those that share roughly the same latitudes with China mostly do not match because of no complicated ecosystem resulted from giant altitude difference, and only a few among those that situated at latitude lower than China exceed the latter. In the four previous books, I have briefed upon the highly complicated and diversified ecosystems of the forests, wetlands and deserts that China boasts, all of which would naturally lead to a huge diversity of wild animals whose population thrive across the vast territory of the country.

Zoogeographical Zoning

Zoogeographical zoning is a prerequisite for a clear and well-organized presentation of the huge diversity of wild animals that find their shelters in China. According to the global zoogeographical zoning, China is often known for bestriding the Palaearctic and Oriental biogeographic realms, whose dividing line extends roughly along the Qinling Mountains-Huaihe River and the south-east boundary of the Qinghai-Tibet Plateau and each of which covers about half of the country's land territory. Mr. Zhang Rongzu drew on this zoning and further divided China into seven biogeographic regions (see *China's Zoogeography* published in 2011), among which the North-east Region (NER), the North China Region (NCR), the Inner Mongolia and Xinjiang Region (IMXR) and the Qinghai-Tibet Region (QTR) fall under the Palaearctic Realm, and the South-west Region (SWR), the Central China Region (CCR) and the South China Region (SCR) under the Oriental Realm. Nevertheless, as a consequence of the highly complicated nature of the country's geographic topography, its hugely variegated natural ecosystems and the great diversity of its wild animal species, as well as of the limit of resources and technologies available back then that resulted in insufficiencies or even total absence of corresponding data, the zoning was rather rough and precise delineation of the boundaries between the seven regions was yet to be optimized.

The rapid social and economic development in China raises an urgent need for the country to take reliable base line information of the wild animal resources that it has, so as to meet the need of conservation and management in this field. In response to this urgent need, China kicked off its first nationwide inventory of terrestrial wild animals in 1995−2003, during which I worked as the chief technical director. A second nationwide inventory of terrestrial wild animals followed in 2011. Such central-planned massive inventory, carried out under the same technical standards and within the same time frame, has never been done by any other major countries across the world. The two large-scale inventories turned out a vast amount of detailed data concerning the distribution of wild animals in the country and laid down a solid foundation for modifying and optimizing the previous zoning of the zoogeographic regions. According to the latest revised *National Zoning of Terrestrial Zoogeographic Regions*, the land territory of the country is divided into seven regions, which break further down to 18 sub-regions that are composed of 239 geo-units in 54 zoogeographic provinces. Major modifications include: a comparatively larger part of the Qaidam

Basin is included in the IMXR; taking into consideration the existence of north-south mountain ridges and deep valleys in the Hengduan Mountains, fine modifications are made to the dividing lines between the QTR and the SWR as well as that between the SWR and the SCR, replacing the rough-made straight borderlines between them with more precisely defined zigzagging lines in between; the south-eastern parts of Tibet, which was previously regarded as a composing section of the QTR, now falls within the SWR, *etc*. In summary, the seven-region zoning scheme adopted in this present album is based on the zoning approach proposed by Mr. Zhang Rongzu as well as on the reliable databank generated from the two above-mentioned nationwide inventories of terrestrial wild animal.

Zoning of marine zoogeography is an exception given that no such zoning whatsoever is yet available in China. For the sake of convenience, a relatively simplified approach is adopted in present album. For the four regions among the seven, namely the NER, the NCR, the CCR and the SCR, the borderlines are extended outwards to the ocean to have the wild marine animal species living there included in the corresponding regions.

About Terrestrial and Aquatic Wild Animals

Traditional approaches in China, including those adopted for the purposes of administration and scientific research and monitoring, break down the wild animals of the country into two large categories: terrestrial wild animals and aquatic wild animals. The present book, taking into consideration the fact that both categories often share the same eco-region, will break away from this categorization and discard the traditional artificially made boundaries. Be they land-dwelling or water-borne, all wild animals that inhabit a certain region among the seven major regions will be presented and introduced together. As a matter of fact, many wild animals inhabiting the same eco-region are often mutually dependent and can hardly be separated from each other. More importantly, in some cases, it is just neither possible nor appropriate to have an artificial clear-cut line concerning the question whether a given species is terrestrial or aquatic, especially amphibian and reptile species.

Basic Principles of Text Writing and Photo Selection

Being the 5th book in the *A Single Drop of Water Ecological Photography Album Series*, the present album will follow the same guiding principle that has been used for the previous ones, that is: telling Chinese eco-stories through the language of photographing. Specifically speaking, it is hoped that a miniature story about a given creature can be told by one single photo, two or three photos make a bigger story about the animal, and a set of photos compose a thorough narration of a given species. Each chapter, which will be dedicated to presenting a holistic profile of a certain zoogeographic region, will add up to form a comprehensive, concise and compelling saga of the rich diversity of wild animals in China.

For sure, this is just the goal that I had committed myself, the journey towards the attainment of which may inevitably fall short of expectation somewhere along the way. While composing this album, I have often found myself troubled by certain puzzling issues. Take for instance, besides the vivid and lifelike pictures, captivating titles and concise texts are also an essential part of an ecological photography album to give the readers as much information as possible about a particular creature or a species. Here come the questions: How can the limited page spaces be used to pack in all the necessary messages, the name (the common name as well as the Latin name), the extent to which it is endangered or threatened, its position in the eco-chain, writer's reflections, place of photographing and so on and so forth? Should a certain species, which is neither eye-appealing nor aesthetically noteworthy but of significant value to the integrity of the eco-system or is capable of making up a compelling, interesting story, be included in the book? What proportion of the album, in terms of both the photos and corresponding explanatory texts, should be devoted to a given species? What should be the proper ratio between the full stories and the miniature ones? All these are but a few examples of the problems that have bothered me throughout the composing process.

Another challenge for me is that, unlike the four previous ones, this album on wild animals must be capable of providing a detailed, concise and accurate introduction about each of the species covered, yet at the same time, as the name implies, an eco-album is different by nature from an illustrated handbook that themes on wild animals. Due to the limits of space, only a single photo can be chosen for a given species. That the photo chosen must be not only representative of the species but also story-worthy means that tradeoffs will often have to be made in the selection process. On some occasions, photos that are of better aesthetical qualities have to be dismissed in favor of making better stories. In a word, there are just too many issues that have to be taken into account in the making of an eco-album that features the hugely diversified wild animals of China, which also explains why the book has taken three long and painful years in coming by, not to mention the decades that have been spent in the wild for the photographing. The photos finally adopted in this book are selected from hundreds of thousands of pictures and most of them are taken in the wild, with the exception of a few featuring tigers, bears and some marine animals that are sheerly impossible to be taken under such conditions. In such rare cases, photos taken at rescue centers, wildlife parks and marine parks (with the specific places clearly indicated) are used, but the total number of these photos does not exceed ten.

There are no previous examples that we can borrow from to create within a book of approximately 200 pages a compelling narration about wild animals in China that not only meets the multiple demands of thoughts, science and art, but also reflects rich diversity. For this reason, detours, trials and errors in the selection of most ideal photos have almost become a part of the process. As a means of artistically presenting something that is expected to be both science-based and thought-inspiring, a good ecological photography album should neither favor the needs of aesthetics at the cost of science nor vice versa, nor should the author shun the challenges and problems that still exist in the field of wild animals protection in China. It is expected to highlight the mutually complementing relation and integration between human beings and nature, while at the same time putting in the author's personal experiences and his reflections concerning the ecosystems. I undoubtedly hope that the multiple and sometimes mutually conflicting needs can be simultaneously met, but reality reminds me that this is a goal that is very hard to reach. Frequent tradeoffs, hesitations, repeated modifications have often frustrated me and made me keenly aware of how insufficient I am for this job.

Ecological Civilization — Building a Shared Future for All Life on the Earth

An overriding principle that has been consistently observed to in composing the albums under *A Single Drop of Water Ecological Photography Series*, the four previous ones featuring various ecosystems and the present one featuring animal species, is that eco-photographing that combines lifelike photos and concise yet compelling texts shall be employed as the means to inform the readers and at the same time to celebrate the vision that takes the Earth as a community for life shared by both human beings and wild animals. To enhance people's awareness for wildlife conservation, this album, while generally maintaining an affirmative tone concerning the achievements that have been made in this regard, does not evade the problems and challenges we are still faced with. In addition, I draw on my personally experience and puts forward my thoughts, making the album a comprehensive and objective narration about the wild animals in China. In a sense, this album might as well be taken as a fairly accessible reading of popular science and will play a role in further improving the understanding of ecological civilization of all nationals in the country.

In July 2019, the International Union for Conservation of Nature (IUCN) released its latest *Red List of Threatened Species*, together with a comprehensive review about the current state of the conservation of endangered species of wild fauna and flora in the global communities. 7,000 species are added to the revised *Red List of Threatened Species*, making the total number of listed species exceeding 100,000 for the first time, with 28,338 species on the verge of extinction. IUCN stresses that, as a result of worsening natural environment caused by population growth, economic development and global warming, as well as unsustainable agriculture and fishery, fragile fauna and flora species are faced with increasingly severe competitions for existence. A latest science report finds that at present, about 1/4 of the fauna and flora species are threatened, with approximately one million species likely to go extinct in the next few decades. Some scientists even claim that, judging from the ongoing speed of species loss, life on the Earth is highly likely to be subjected to massive bio-extinction for the sixth time.

The United Nations *Convention on Biological Diversity*, a legally-binding convention with 196 signatory countries, aims to protect the endangered fauna and flora species so as to conserve the diverse biological resources on this planet to the greatest extent for the benefits of the current and future generations. The COP15 of the United Nations *Convention on Biological Diversity* that will soon be coming up in Kunming identifies its theme as "Ecological Civilization — Building a Shared Future for All Life on the Earth". This conference will be of great significance for mobilizing the political will of the international community for biodiversity conservation to promote the progress of ecological civilization globally, for advancing the objectives of the CBD of sustainably using and conserving the biodiversity and achieving the vision of "living in harmony with nature" by 2050. It is also anticipated that the conference will adopt the *Post-2020 Global Biodiversity Framework*, urging the global communities and member countries to take necessary measures and steps to curb the trend of biodiversity loss, and that new biodiversity conservation goals for 2030 will be identified.

China, one of the countries with the greatest biodiversity in the world, has taken initiative in putting forward the comprehensive treatment of mountains, rivers, forests, farmlands, lakes, grasslands and deserts as well as the idea of green development and delimitating red lines for ecological conservation, with over 18% of its territory having been designated as protected areas of various types, where a huge quantity of natural ecosystems and the majority of critical wild animal species are protected. Moreover, China will strive for the new goal of global biodiversity in 2030. The theme of the upcoming conference, "Ecological Civilization — Building a Shared Future for All Life on Earth", resonates with the global trend for green development and voices the common aspirations of people across the world. It is a vision that we Chinese pledge our unwavering commitment. It is also the reason that inspired me to compose this album.

林海雪原·吉林延吉
Snow-covered forests
Yanji, Jilin

本区紧靠东西伯利亚，夏季温暖短暂，冬季寒冷漫长，是中国最冷和降雪最多的地方。这里地表积雪时间长久，使温带森林呈现出一派林海雪原的美丽景观。
Adjoining the eastern Siberia, the NER has short warm summer but long harsh winter, thus making it one of the coldest places in China with the greatest amount of snow. As the land is covered by snow for a long period of time, the temperate forests in this region often present us beautiful landscape of icy sea of forests.

东北 North-east Region 区

生物宝库·吉林长白山
The treasure-house of biodiversity — Changbai Mountains, Jilin

整个长白山森林生态系统是我国北方最珍贵、最特有的生物多样性宝库，也是野生动物的乐园。
The entire forest ecosystem in Changbai Mountains is the most precious and unique bank of biodiversity as well as a natural paradise for wild animals.

典型林带·吉林抚松
Representative forest belts — Fusong, Jilin

白桦林、红松阔叶混交林、长白落叶松林是本区中温带森林生态系统的典型代表。
Birch (*Betula platyphylla*) forest, mixed Korean pine (*Pinus koraiensis*) and broad-leaved forest as well as Korean larch (*Larix olgensis*) forest make up the defining landscape in the forest ecosystems of the NER situated in the middle temperate zone.

东北大湿地·黑龙江南瓮河
Wetlands in the northeastern China — Nanweng River, Heilongjiang

在东北区的永冻层上，河流舒缓，千回百转，湿地广阔，鸟类翱翔。
On the permafrost of the NER, rivers meander slowly across vast wetlands, providing a pleasant home to the rich variety of birds soaring carefree in the sky overhead.

东北区 包括大兴安岭、小兴安岭、张广才岭、老爷岭、长白山地、松嫩平原、辽河平原及黄海和渤海东部,属世界动物地理区划中古北界的中国东北部分,地处寒温带和中温带,气候寒冷严酷。动物区系成分主要为古北型、全北型及东北型,动物种类相对贫乏(尤其是两栖爬行类),但生存着一些大型哺乳类动物如东北虎、东北豹、棕熊、黑熊、梅花鹿、马鹿、狍等,一些大型鸟类春秋两季迁徙经过这里,如丹顶鹤、白鹤、白枕鹤及白头鹤等,常见雕鸮、乌林鸮、长尾林鸮、白尾海雕和雪雕等猛禽。不少哺乳动物有冬眠现象或储藏食物越冬的习性,而两栖动物、爬行动物和大量低等动物都有冬眠的现象。松嫩平原的大量湿地及辽河口、鸭绿江口湿地是多种鹤、鸻鹬类鸟每年大批迁徙过程中停歇和补充营养的地方,是世界著名东亚—澳大利西亚迁徙路线上非常重要的驿站。

"棒打狍子瓢舀鱼,野鸡飞到饭锅里",这里曾经是野生动物非常丰富的地方。但是,20世纪以来原始森林遭到大量采伐,人口增加带来了大规模开荒开湿种地,导致了野生动物的栖息地受到严重大破坏甚至丧失,野生动物数量大大减少。该地区最有代表性的顶级物种——东北虎、东北豹也几乎绝迹,仅在靠近俄罗斯的边境地区偶尔发现有游荡虎和零星虎、豹。20世纪末以来,国家实施了天然林资源保护、野生动植物保护和自然保护区建设、湿地保护、退耕还林还湿等生态建设工程,该地区完成了从减少森林采伐量逐步过渡到现今的全面停止采伐,从部分禁猎到全面禁猎的历史性转变。另外,为了扩大野生鲟鳇鱼等鱼类的种群,还对黑龙江等江河里的鱼类进行了保护和增殖放流等。这些一系列政策的实施,从根本上扭转了该地区自然生态环境恶化的局面,使野生动物无论从种类还是数量上都得到了较大恢复。

该地区东部森林是中国唯一的东北虎、东北豹的栖息地。这些年来,随着自然保护力度的加大以及"东北虎豹国家公园(试点)"(2017年)的建立,虎、豹食物链上的鹿、狍、野猪等野生动物数量得到稳步增长,东北虎、豹的数量扭转了下降的趋势而稳中有升,就连小兴安岭腹地也出现了野生虎的踪迹。东北虎、豹这些伞护种的保护大大促进了东北自然生态系统的恢复完善。在此基础上,建议将建设大兴安岭、长白山国家公园列入议程,为整个东北区自然生态系统的恢复和生物多样性的保护打下更加坚实稳固的基础。

当今,"以国家公园为主体的自然保护地体系"建设的加快必将大大促进自然生态系统不断朝着良性循环的方向转化,该地区野生动物种类和种群的恢复大有希望!

The North-east Region (NER) is located in the cold temperate and middle temperate zones, where the climate is cold and harsh. It includes the Greater Khingan Mountains, the Lesser Khingan Mountains, Zhangguangcai Mountain, Laoye Mountain, Changbai Mountain, Songnen Plain, Liaohe River Plain and the eastern part of Yellow Sea and Bohai Sea. This region locates in north-eastern China, which falls under the Palaearctic Realm in world zoogeographic zoning. The fauna in this region mainly belong to one of the following three zoogeographic realms: Palearctic, Holarctic and the Northeast. Wild animal species inhabiting the NER is comparatively less diversified, amphibians and reptiles in particular. Nevertheless, some large mammals live here, such as Siberian tiger (*Panthera tigris altaica*), Amur leopard (*Panthera pardus orientalis*), brown bear (*Ursus arctos*), black bear (*Ursus thibetanus*), sika deer (*Cervus hortulorum*), Machurian wapiti (*Cervus canadensis*) and roe deer (*Capreolus pygargus*), *etc*. Some large birds migrate through here in spring and autumn, such as red-crowned cranes (*Grus japonensis*), white cranes (*Grus leucogeranus*), white-naped cranes (*Grus vipio*) and hooded cranes (*Grus monacha*). Eagle owls (*Bubo bubo*), Ural owl (*Strix nebulosa*), brown wood-owl (*Strix uralensis*), white-tailed sea eagles (*Haliaeetus albicilla*) and snow eagles can be easily found here. Quite a few mammals hibernate or have the habits of overwintering with the food storage. And amphibians, reptiles and a large number of lower animals hibernate. Plenty of wetlands within the Songnen Plain and the river mouth of Liaohe River and Yalu River, where many kinds of cranes and waders stop temporarily during their annual long-distance migration for rest and food, serve as important posthouses on the famous migration route of East Asian-Australasian Flyway (EAAF).

A saying that used to be heard often among the local residents goes that "so rich are the wild animals here that you can club a roe deer easily, scoop up fish from the rivers at any time. You can even just wait for an unwitting pheasant to fly into your pot." It used to be a place of huge diversity of wild animals. Since the beginning of the 20th century, however, deforestation of old-growth forests, massive reclamation of wasteland/wetland for farming have led to severe degradation and even total loss of habitats of wild animals, leading to an alarming decline in wildlife population. Take the Siberian tiger and Amur leopard, the two most representative top species in this region for example, they have almost gone extinct, with just a couple of lonely tigers and leopards found occasionally near the border between China and Russia. Thanks to the implementation of a series of key national ecological rehabilitation programs since the late 20th century — including the Natural Forest Protection Program (NFPP), the National Wildlife Protection and Nature Reserve Development Program, the Wetlands Conservation Program, and the Conversion of Cropland to Forest/Wetland Program, the region has witnessed the shift from a phase marked by gradual decrease in logging, partial ban on hunting to a phase when a total ban is placed on all logging and hunting activities. In addition, measures have been taken to protect and boost the population of fish like wild bastard sturgeon (*Acipenser nudiventris*) and other species in the Heilongjiang River and other water systems. The implementation of such measures has reversed the deteriorating trend of the natural ecological environment in this region that contributed greatly to the recovery in both the species and population of wild animals.

Forests in the eastern part of the region are the only home to the Siberian tiger and Amur leopard in China. With the increased efforts in nature conservation, and following the establishment of the National Park (Pilot) for Siberian Tiger and Amur Leopard in 2017, the number of wild animals such as deer, roe deer and wild boars in the food chains of tigers and leopards have increased steadily. The populations of Siberian tigers and leopards have reversed their declining trend and climbed up steadily, with sight of wild tigers reported even at the heartlands of the Lesser Khingan Mountains. Effective protection of the umbrella species like Siberian tigers and Amur leopards greatly promoted the restoration and improvement of the ecosystem in the northeastern China. Drawing on the success that have been made, we suggest that the establishment of the Greater Khingan Mountains National Park and the Changbai Mountain National Park should be put on agenda as soon as possible, so as to lay a more solid foundation for the restoration of the entire ecosystem and the conservation of biodiversity in NER.

Acceleration in the development of a natural protected area system highlighting the central role that national parks play will greatly promote the transformation of the natural ecosystem to a virtuous circle. A bright future can be expected for this region in the recovery of wildlife species and corresponding populations.

↑ 交配结束·黑龙江虎林园
Tigers after mating — Tiger Park in Heilongjiang

东北虎在冬季交配，在完成交配的瞬间，雌老虎为了摆脱疼痛，会猛然转过头来，咬退身上的雄虎。
Siberian tigers (*Panthera tigris altaica*) mate in winter. Once they have completed that process, the female tiger would suddenly turned around to bite the male tiger off her back in order to ease the pain.

中国老虎保护的希望之地
The Land of Hope for Tiger Protection in China

虎起源于黄河中游一带，随后逐步扩散到亚洲其他国家，成为亚洲的特有物种。虎共分化出8个亚种，其中，巴厘虎、里海虎、爪哇虎现已灭绝，现存5个亚种分别为东北虎、印度支那虎、孟加拉虎、华南虎和苏门答腊虎，中国有除了苏门答腊虎之外的另外4种虎的分布。老虎处于自然界食物链的顶端，是生态系统的旗舰物种，具有重要的科学、生态、美学、文化价值。中国的虎文化源远流长，崇虎敬虎的思想无处不在，降虎伏虎的故事多有流传，有如调兵遣将的虎符，虎头鞋帽的民俗，武松打虎、打虎上山的故事，十二生肖中的虎属等。

东北虎历史上分布相当广泛，包括中国华北北部、东北大部、朝鲜半岛、蒙古国东北部及俄罗斯远东大部。据科学估计，19世纪末全世界东北虎的总数有2000~3000只，中国大约占一半多，后至20世纪初，东北虎的分布范围逐渐缩小。20世纪50年代以后，由于中国捕虎运动及林区人口数量激增，东北虎被迫迁移到干扰较少的边境地区，仅生存于完达山、乌苏里江畔及珲春、汪清一带。1998—1999年，中国、俄罗斯、美国三国专家联合调查结果显示，中国的东北虎仅存12~16只、东北豹7~12只。

改革开放之后，中国政府加大了对野生动物保护的力度，先后建立了若干个以老虎为主要保护对象的自然保护区。我考察过多个虎自然保护区，也曾受国际野生生物保护学会（WCS）的邀请，到俄罗斯远东的阿林虎自然保护区、豹乡自然保护区等地考察。通过考察，我了解到，在俄罗斯政府的积极努力下，目前俄罗斯的老虎从50年前的30~40只增长到创纪录的400~500只，他们的很多经验值得我们学习。世界自然基金会（WWF）将2010年定为"国际老虎年"，该年9月，在俄罗斯召开的包括中国在内的世界老虎主要分布国及国际组织参加的"全球老虎峰会"上提出，老虎主要分布国要作出努力，力争在2022年再逢"虎"年时，全球野生老虎数量能够翻一番。

目前，中国的华南虎、印度支那虎、孟加拉虎数量已经极为稀少，尽管做了大量艰苦的努力，但野外种群恢复困难极大，只有东北虎的种群恢复已经向着良好的方向转化。随着中国天然林资源保护工程和退耕还林还草工程的深入实施，特别是吉林、黑龙江两省20世纪90年代中期开始实施的全面禁猎及之后的全面禁伐，东北虎豹栖息地得到了大大改善，野生种群得到了逐步恢复。2012—2014年，东北虎已恢复到27只，东北豹恢复到42只。2017年8月，地跨吉林、黑龙江两省，总面积146万公顷的"东北虎豹国家公园"挂牌，迎来了中国东北虎豹保护的新纪元。最新数据显示，国家公园试点至今，已记录到新繁殖的东北虎幼崽12只，东北豹幼崽11只以上。

东北虎的拯救是中国野生动物保护工作中在保护力度和种群数量方面都取得明显提高的典型案例，中国老虎恢复的希望于此。2022年再逢"虎"年，中国野生老虎总体数量比上一个虎年翻一番的目标是肯定能够实现的。

Tigers originated in the middle reaches of the Yellow River, and then gradually spread its populations to other nations in Asia and became the endemic species of the continent. There are eight subspecies of tiger, among which the Bali tiger (*Panthera tigris balica*), the Caspian tiger (*Panthera tigris virgata*) and the Javan tiger (*Panthera tigris sondaica*) have gone extinct. The five remaining subspecies are the Siberian tiger (*Panthera tigris altaica*), the Indochinese tiger (*Panthera tigris corbetti*), the Bengal tiger (*Panthera tigris tigris*), the South China tiger (*Panthera tigris amoyensis*) and the Sumatran tiger (*Panthera tigris sumatrae*). Except the Sumatran tiger, all the other 4 subspecies exist in China. Standing on the very top of food chains, tigers are considered as flagship species for its scientific, ecological, aesthetic and cultural values, which are of great importance to our ecosystem. Tiger culture in China goes back to time immemorial, which can be fully proved by both the ubiquitous ideas of tiger worshiping and legendary folk stories about how human beings conquer and subdue tigers. The tiger tallies awarded to army generals in ancient China as symbols of their commanding power over the military troops, the tiger-head-shaped shoes and hats often seen in traditional Chinese folk handicraft, the legendary household stories like *Wu Song Slaying the Tiger* (a chapter taken from one of China's the four most-favored literary classics) and *Subduing the Tiger before Joining the Bandits* (a most popular Peking opera sketch), and the Tiger in Chinese Zodiac, all these make up telling examples about the popularity that tiger enjoys among the Chinese people.

According to historical records, Siberian tigers used to be

extensively distributed over vast areas, including the northern parts of North China, most parts of northeast China, the Korean Peninsula, the northeastern part of Mongolia and the greater part of the Russian Far East region. It is estimated that, at the end of the 19th century, approximately half of the world's total population of Siberian tigers, which stood roughly at 2,000-3,000, lived in China. By the turn of the 20th century, the distribution area of this species had already notably shrunk. As a result of China's tiger hunting campaign in the 1950s and population growth in the forested areas in the decades that followed, suitable habitats of Siberian tigers receded gradually to border areas where disturbances caused by humans are less. In other words, they currently range only a narrow stretch of lands in areas like the Wanda Mountains, the Ussuri River valley, and in counties like Hunchun, Wangqing and *etc*. In 1998-1999, according to a joint investigation conducted by experts from China, Russia and the United States, just 12-16 Siberian tigers and 7-12 Amur leopards still exist in China.

Since the reform and opening-up, the Chinese government has stepped up efforts to protect wildlife and set up a number of nature reserves that feature tigers as the chief targets of protection. I have been to many such nature reserves. At the invitation of the Wildlife Conservation Society (WCS), I have also been to Sikhote-Alin Nature Reserve and the Land of the Leopard National Park in the Far East area of Russia for field investigations concerning tiger protection. Through these experiences, I learned that through the vigorous efforts taken by the Russian government over the past 50 years, tiger population in Russia has increased from 30-40 to a record high of 400-500, Many of their practices and expertise are well worthy for us to learn from. WWF designated 2010 as the International Year of Tiger, and at the Global Tiger Summit held in Russia in September that year, participating countries and international organizations jointly called on all tiger-inhabiting countries to step up their protective efforts so that the global population of wild tigers could double by 2022, the next Year of Tiger in Chinese zodiac.

At present, the populations of South China tigers, Indochinese tigers and Bengal tigers are extremely small in China. Despite

虎视眈眈·
黑龙江横道河子
Eyeing its prey —
Hengdaohezi, Heilongjiang

雪中卧虎不减雄风，虎视眈眈伺机而动。
In the snow is a crouching tiger with undiminished muscular strength, waiting for the right time to jump at its prey.

the strenuous efforts that have been made, it is extremely difficult to restore their wild population. In fact, promising signs have been noticed in only the protective efforts of Siberian tigers, whose population is on the trend of recovering. With the further implementation of the Natural Forest Protection Program (NFPP) and the Conversion of Cropland to Forest/Grassland Program (CCFGP) in China and the total ban on hunting in Jilin and Heilongjiang provinces since the mid-1990s and the total ban on harvesting thereafter in particular, the habitats of Siberian tigers and Amur leopards have been greatly improved, accompanied by notable recovery in their wild population. During 2012-2014, the populations of Siberian tigers and Amur leopards in China had recovered to 27 and 42 respectively. In August 2017, the 1.46 million-hectare Siberian Tiger and Amur Leopard National Park, which covers the areas in both Jilin and Heilongjiang provinces, was officially established, ushering in a new era for the conservation for Siberian tigers and Amur leopards in China. According to the most updated data, more than 12 new born Sibirian tiger calves and 11 new born Amur leopard calves were recorded since the piloting of National Parks.

The rescue of Siberian tigers is a telling example that showcases the great improvements that China has achieved both in strengthening its protective efforts and in recovering wildlife populations in the country. This marks the very area where the hope and future of tiger protection lie. We have full confidence that the goal to have the population of wild tigers in China doubled (as compared with that of the previous Year of Tiger) by 2022 will be reached.

鹤类种数最多的国家
The Country with the Richest Crane Diversity

现今世界15种鹤中有9种分布于中国，中国是世界上鹤类种数最多的国家。鹤在中国传统文化中具有崇高的地位，尤其丹顶鹤因常与神仙联系起来，被称为"仙鹤"。唐宋年间，咏鹤诗词数量众多并一直流传至今，明清时期一品文官服上才有资格绣上仙鹤图案。鹤被中华文化赋予美丽、飘逸、长寿、吉祥和高贵的寓意，已经成为人文精神高洁的象征，常见有寓意幸福长寿的"松鹤延年图"等。

鹤类为大型迁徙鸟类，喜欢结群生活。除黑颈鹤生活繁殖在青藏、云贵高原，赤颈鹤生活在云南南部，沙丘鹤极少量出现在华北区一带以外，其余鹤类繁殖均在北方甚至更远，大致每年10月下旬左右经东北区、蒙新区迁至长江流域一带（蓑羽鹤和部分灰鹤飞得更南，甚至翻越喜马拉雅山脉）越冬，次年4月再到或经东北区、蒙新区飞回北方，白鹤繁殖地甚至到达北极圈附近。

我国的9种鹤中，最漂亮、最著名、最有文化故事的是丹顶鹤；世界几乎全部种群都在鄱阳湖越冬的是白鹤；数量最多、分布最广的是灰鹤；仅仅分布在青藏、云贵高原且迁徙路途较短，也是被发现得最晚的是黑颈鹤；体型最小但却能够飞越喜马拉雅山脉越冬的是蓑羽鹤；有时与灰鹤在同一地区营巢，又共同迁徙且偶有杂交的是白头鹤；个头与白头鹤差不多，繁殖区、越冬区也大致相同的是白枕鹤；最难得一见的是沙丘鹤；生境最南，个体最高大、最强壮，几乎见不着的是赤颈鹤。

Among the 15 crane species existing in the world, nine are found in China, making it the country with the richest crane diversity. Being of high status in Chinese traditional culture, the crane, the red-crowned crane in particular, is called the "fairy crane". During Tang (618–907) and Song (960–1279) dynasties, a large number of poems depicted the fond emotions that the Chinese have since ancient times held towards cranes. During Ming (1368–1644) and Qing (1636–1912) dynasties, only top-level civil servants in feudal regime had the privilege to wear official robes with embroider patterns of the Fairy Crane. The implication of beauty, elegance, longevity, auspiciousness and nobleness given by the Chinese culture to the crane has become a symbol of humanity and noble spirit. For instance, paintings featuring pine and crane are usually understood as signifying longevity.

Cranes are large migratory birds that prefer to live in groups. With the exception of black-necked cranes (Grus nigricollis), which breed in the Qinghai-Tibet Plateau and the Yunnan-Guizhou Plateau, red-necked cranes (Grus antigone), which live in the southern Yunnan, and sandhill cranes (Grus canadensis), which are occasionally found in North China, all other cranes breed in areas further to the north. Each year in late October, they would migrate to the Yangtze River Basin via the Northeast Region (NER) and the Inner Mongolia and Xinjiang Region (IMXR) for overwintering, among which demoiselle cranes (Grus virgo) and some common cranes (Grus grus) would fly further south and even over the Himalayas and in the following April, they would fly back to the NER and IMXR regions or fly further northward from there. The breeding place of white cranes (Grus leucogeranus) extends as far as to the Arctic Circle.

Of all the nine crane species inhabiting China, the red-crowned crane is the most beautiful, the most famous, and the most richly loaded with cultural implications; almost the entire population of white cranes in the world overwinter in the Poyang Lake; common cranes are the most abundant and the most widely distributed; black-necked cranes, which are found only in the Qinghai-Tibet Plateau and the Yunnan-Guizhou Plateau and with a relatively short migration route, are the latest to be discovered; demoiselle cranes are the smallest in size, but they are capable of flying over the Himalayas to overwinter; hooded cranes (Grus monacha) sometimes build nests in the same place as common cranes do and occasionally interbreed with the latter during their shared migratory journeys; white-naped cranes (Grus vipio), which are about the same size as hooded cranes, share roughly the same breeding places and wintering regions as the latter; sandhill cranes are the most difficult to be seen; the southernmost-dwelling Sarus crane (Grus antigone) is the tallest and strongest type among all cranes, but it is almost a once-in-a-blue-moon event for people to actually catch sight of it.

↑ 母与子·内蒙古赤峰
Mother and child — Chifeng, Inner Mongolia

小白枕鹤跟着妈妈逐渐长大了，练练翅膀，毕竟还要离开这里飞往几百千米以外的越冬地。
The little white-naped crane (Grus vipio) gradually grows up with its mother. Now is time to exercise its wings, for it has no choice but to leave here and fly across several hundred kilometers away to overwinter.

← 鹤的一家·黑龙江扎龙
A crane family — Zhalong, Heilongjiang

松嫩平原的湿地是最佳的鹤类迁徙停歇之地。图为日出东方红，温馨的丹顶鹤一家。
The wetlands within the Songnen Plain serve as the best resting place for cranes during their migration. In this picture, a warm family of red-crowned cranes are bathed at the dawn of rising sun.

↑ 白鹤迁飞·吉林向海
Migrating white cranes flying over the sky — Xianghai, Jilin

秋末冬临，各路白鹤迁徙大军集群在蓝天里，飞往的几乎都是同一个目标，即中国南方的越冬圣地——鄱阳湖。世界上98%的白鹤种群都在那里越冬，并成为鄱阳湖湿地生态系统的指示物种。

In the late autumn and early winter, white cranes (*Grus leucogeranus*) coming from all directions meet together in the blue sky. And almost all of them share the same destination, the Poyang Lake, a place in the southern China where they can live through the winter. It's home to 98% of the world's population of white cranes, which has become the indicator of the Poyang Lake wetland ecosystem.

雨季中的蓑羽鹤·内蒙古赤峰
Demoiselle cranes in the rainy season — Chifeng, Inner Mongolia

雨季的草原格外地绿，大量的水泡子给蓑羽鹤提供了更多的栖息地。

Grasslands are extraordinarily green during the rainy season and ponds formed after rain everywhere provide more habitats for the demoiselle cranes (Grus virgo) to live on.

伺机而动·内蒙古兴安盟
Waiting for action — Hinggan League, Inner Mongolia

低飞的白腹鹞并非对躲在芦苇丛中谈情说爱的凤头䴙䴘感兴趣，它已经看见了猎捕对象，伸长了利爪，伺机而动。

The eastern marsh harrier (Circus spilonotus) flying low in the air is not interested in the two great crested grebes (Podiceps cristatus) that are courting in the reeds. Instead, it has already locked its attention on its prey and is waiting for the perfect time to launch its attack with the outstretched sharp claws.

灰鹤迁徙·辽宁朝阳
Common cranes in migration — Chaoyang, Liaoning

灰鹤，别名"千岁鹤"，一种典型的长距离迁徙鸟类，几乎整个亚洲的繁殖种群都在俄罗斯，越冬地延伸到我国东北部和新疆北部，有的甚至到东南亚。

The common crane (Grus grus), alias Qian Sui He (1,000-year-old crane) in Chinese, is a kind of typical birds who can travel long-distance for migration. Almost all the common crane flocks in Asia live in Russia for breeding, with their winter habitats extending to the northeastern China and the northern Xinjiang, and sometimes even as far as Southeast Asia.

← 滑翔捕食·内蒙古呼伦贝尔
Gliding snowy owl is hunting for food —
Hulunbuir, Inner Mongolia

雪鸮在北极和西伯利亚繁殖，越冬时也会出现在中国的寒温带——大兴安岭北部地区，但十分罕见。
Snowy owls (*Bubo scandiacus*) breed in the Arctic and Siberia. They are sometimes found (but on very rare occasions) in the northern Greater Khingan, which falls under the cold temperate climate zone, to live through the winter.

① 似睡非睡·黑龙江加格达奇
Sleeping or not? —
Jiagedaqi, Heilongjiang

乌林鸮在中国仅栖息于大兴安岭原始针叶林和针阔混交林中，面部有呈波状的黑色同心圆是其特点，飞翔迅速无声，常单独停息在高大乔木上，静静地等待和观察猎物。
Great grey owls (*Strix nebulosa*) only inhabit the primary coniferous forests and the mixed coniferous-broad leaved forests of the Greater Khingan. With black concentric circles on their faces as the distinctive features, they fly quickly and silently and often rest alone in tall trees, keeping close track of their preys.

② 睁一只眼闭一只眼·辽宁大连
Seemingly absent-minded —
Dalian, Liaoning

雕鸮分布范围较广，多躲藏在密林中栖息，缩颈闭目立于树上，但它的听觉甚为敏锐，稍有声响，立即伸颈睁眼，转动身体，观察四周动静。
Eurasian eagle-owls (*Bubo bubo*) are extensively distributed over vast areas. On most occasions, they seem to be perched in dense forests, with eyes closed and neck withdrawn, but in fact they are constantly on alert with their acute auditory sense. Any tiny sound in the air would be sufficient to alarm them into vigilance.

雪中猎渔·吉林珲春
Fishing in the snow —
Hunchun, Jilin

白尾海雕为大型猛禽，主要以鱼为食，常在水面低空飞行，抓到鱼后会飞到邻近的树上或高地上食用，即便是大雪纷飞也丝毫不影响它捕食的兴致。

The white-tailed eagle (*Haliaeetus albicilla*) is a kind of large bird of prey, mainly living on fishes and often flying low over water surface. After catching a fish, it will fly up to a tree nearby or a higher land to enjoy their meal. Even snowy days will not bother it in its passion for preying.

◀ 水上之爱·吉林长白山
Love on the water — Changbai Mountains, Jilin

初春季节，在林中清澈的河溪中，世界上最古老的鸭子——中华秋沙鸭相互追逐，不失时机地做爱，为繁殖后代作准备。
In early spring, two Chinese mergansers (*Mergus squamatus*), the oldest ducks in the world, are chasing after each other in the crystal-clear streams in the forest and losing no time to mate and get ready for breeding their offspring.

❶ 妈妈回来了·黑龙江大兴安岭
Mom's back — Greater Khingan, Heilongjiang

在草原的石头崖壁上，黄爪隼雏鸟的个子已经长得和妈妈差不多了，妈妈带回了孩子生长亟需的食物（沙蜥），大家无比兴奋。
The nestling lesser kestrels (*Falco naumanni*) on the rocky cliffs by the grassland are now of almost the same size as their mother who has taken home the much-needed food (*phrynocephalus* sp.) for her children.

❷❸❹ 雪中小鸟·吉林长白山
Birds in the snow — Changbai Mountains, Jinlin

在本区长达半年的冬天里，林海雪原中照样活跃着各种各样的小鸟：黑头蜡嘴雀东北亚种（图2）、普通鸭（图3）、银喉长尾山雀（图4）。
During the almost-half-year long winter in the NER, snow-covered forests are still teeming with birds of a huge variety, including the magnirostris of Japanese grosbeak (*Eophona personata magnirostris*, Photo 2), Eurasian nuthatch (*Sitta europaea*, Photo 3), and long-tailed tit (*Aegithalos glaucogularis*, Photo 4).

↑ 多种多样的兽类·东北区
A great diversity of beasts — NER

泡子中的两狍·内蒙古额尔古纳 →
Two roe deer in a pond — Ergun, Inner Mongolia

本区的森林、草原、湿地中同样活跃着很多大、小型兽类，例如，岩松鼠（图1）、紫貂（图2）、驯鹿（图3）。

Forests, grasslands and wetlands in this region are also the great paradise for a great deal of large and small mammals like the David's rock squirrel (*Sciurotamias davidianus*, Photo 1), Japanese sable (*Martes zibellina*, Photo 2), and reindeer (*Rangifer tarandus*, Photo 3).

狍（俗称狍子）是东北地区常见的野生动物之一，原东北俗语"棒打狍子"指的就是它。狍性情胆小，日间多栖于密林中，早晚时分才会在空旷的草场或灌木丛中活动，狍子在水泡中活动实为罕见，如果不是航拍，很难发现这种情况。

The roe deer (*Capreolus pygargus*), known as Pao Zi in local dialect, is one of the wild animals commonly seen in the NER. It is so widely seen that "to club a roe deer" has become a household phrase popular among the local folks. The roe deer is the timid animal that hides deep in forests during the day and can be seen on open grassland or under low bushes only at dawn and dusk. It is indeed a rare event to see the roe deer in the pond. Hadn't it been for the help of aerial photographing technology, such a scene wouldn't have been captured.

呦呦鹿鸣·黑龙江漠河
The bleating of deer — Mohe, Heilongjiang

古人很早就认识了梅花鹿并对它寄托了丰富的情感。"呦呦鹿鸣，食野之苹"，就取之于中国最早的诗歌——《诗经》。

The Chinese have held fond emotions towards sika deer (*Cervus nippon*) since ancient times. "With pleasant sounds the deer bleat at one another, eating the mugwort of the fields at leisure" is a well-known line often quoted from the *Book of Songs*, one of the earliest collections of Chinese poems written in the Pre-Qin Era.

孤独巨兽·长隆野生动物园
Lonely giant beast — Guangzhou Chimelong Safari Park

棕熊没有固定的栖息场所，平时单独行动，食性较杂，有冬眠的习性。

The brown bear (*Ursus arctos*) has no fixed habitat and normally lives in isolation. It feeds on a large variety of food and has the habit of hibernation.

舔食·长隆野生动物园
Licking — Guangzhou Chimelong Safari Park

黑熊是典型的林栖动物，杂食性，春天主要以树芽为食，夏天喜食蚂蚁、蜜蜂等，尤爱舔舐蜂蜜，秋天主要以壳斗科植物的果为食，储存能量进入冬眠。

The black bear (*Ursus thibetanus*) is a typical forest-dwelling animal. Being omnivorous in food, they would feed mainly on the buds of trees in spring, eat ants, bees and other insects with honey as its favorite food in summer, and have a tendency to eat quite a lot fruits grown in trees in the Fagaceae family during autumn seasons so as to store enough energy for hibernation in winter.

◀ 山坡上的回眸·
黑龙江小兴安岭
Looking back on hill —
Lesser Khingan,
Heilongjiang

马鹿生活于高山森林或草原地区。夕阳西下，机警的马鹿回眸一看，和我的镜头正好相对。

The Manchurian wapiti (*Cervus canadensis*) is found in the alpine forests or grasslands. Just as the sun is about to set, the vigilant deer happens to turn her head back, right into the lens of my camera.

生态思考 Ecological Reflection
天然林全面停伐之后
After the Implementation of Complete Logging Ban on Natural Forests

新家·内蒙古牙克石
New home — Yakeshi, Inner Mongolia

在森林全面停伐、林场撤场下山、砍树人成护林人的大背景下，废弃的原林场职工住房变成了长尾林鸮的新家。

Under the circumstance of full logging ban when former forest farms are retreated from the hills, former lumber jacks assume their new roles as forest rangers, houses previously inhabited by the staff of these farms are abandoned and become new homes to the Ural owl (Strix uralensis).

新气象·黑龙江牡丹江
Brand-new scenery — Mudanjiang River, Heilongjiang

在国家重大政策战略转移的指引下，林海雪原中再也没有了往年冬季采伐作业的喧嚣，连集材场都变成了新造林地，呈现出了一派祥和宁静的美丽景象。

Thanks to the strategic shifts in the country's forestry policies, the bustling and hustling logging operations previously prevalent in the snow-covered forests during winter months are no longer seen, even the forest depot become new reforestation land. Instead, peace and tranquility prevail now throughout the beautiful landscape of forests.

天然林是结构最复杂、生态功能最强大、生物多样性最丰富的生态系统，是陆地生态系统主体——森林的精华。中国现有天然林面积占全国森林面积的64%左右，蓄积量占全国森林蓄积量的83%以上。从维护我国生态安全的战略高度看，天然林保护承担着构建中国完备的生态保障体系主体、保护生物多样性、践行"绿水青山就是金山银山"等极为重要的任务。

"天然林资源保护工程"是1998年长江、松花江流域发生特大洪灾后，党中央、国务院作出的一项重大战略决策，是中国林业以木材生产为主向以生态建设为主转变的重要标志。之后，中国政府又于2015年全面停止内蒙古、黑龙江、吉林等重点国有林区的商业性采伐，2017年全面停止全国的天然林商业性采伐。天然林资源保护工程的实施，使被破坏的、退化的森林植被得到较快恢复和重建，工程区水土流失、泥石流、山体滑坡等灾害大为减少，有效地保障了大江大河安澜和国土生态安全。工程实施20年来，国家为天然林资源保护投入资金已达4000多亿元。

东北林区是为新中国建设作出过巨大贡献的老重点林区，由于采取了从减少森林采伐量逐步过渡到全面停止森林采伐，从部分禁猎逐步过渡到全面禁猎等重大工程措施，实现了"林进民退"，很多林场、站点都撤退下山，再没有职工、居民居住，林业工人由"砍树人"变为了"护林人"，得到了妥善的安置和稳定的就业。天然林资源保护工程的实施，为野生动物的生存繁衍提供了良好的环境，工程区内已消失多年的飞禽走兽重新出现，野生动物无论从种类还是数量上都得到了极大恢复。

Natural forests, for the most complex structure, the most powerful ecological functions and the richest biodiversity they have, are critical components of forests that make up the major part of the terrestrial ecosystem on the planet. Natural forests in China account for about 64% of its existing forests and contribute up to 83% of the country's total forest growing stock. Judged from the strategic viewpoint of national ecological security, the protection of natural forests is of the utmost importance to putting in place for China a comprehensive eco-security system, to the conservation of biodiversity, as well as to bringing into reality our cherished vision that "lucid water and lush mountains are invaluable assets".

The "Natural Forest Protection Program" (NFPP) was a major strategic decision made by the CPC Central Committee and the State Council after the devastating floods that hit the basins of the Yangtze River and the Songhuajiang River in 1998. The implementation of it marked a transitional point when China committed itself to moving from the forest development strategy that gave priority to timber-production to one that stresses on the ecological functions of forests. Later in 2015, the Chinese government placed a complete ban on all commercial logging in key state-owned forest areas such as Inner Mongolia, Heilongjiang and Jilin, which was expanded in 2017 to cover all natural forests in the country. The implementation of the NFPP has led to the rapid recovery and rehabilitation of the damaged and degraded forest vegetation and the reduction of water loss and soil erosion, debris flow, landslides and other disasters in the project area, ensuring the peace of rivers and the ecological security of our country. Over the past two decades since the implementation of the NFPP, the state has invested over 400 billion yuan in the protection of natural forest resources.

The forest regions in northeastern China are traditional key

forest regions that have made great contributions to the development of the People's Republic of China. Following the NFPP, which has evolved from gradual reduction in logging quotas to complete harvest ban, from restricted hunting to complete ban on hunting, many forest farms and stations were consecutively closed. Forestry employees, woodland-dwellers that used to inhabit in the forests gradually moved out, achieving the status that "forests encroach while human retreats". Former lumber jacks were re-trained to assume new roles as forest rangers, or repositioned into other sectors for alternative livelihood. Thanks to the implementation of the NFPP, wild animals have been provided a good environment for their survival and reproduction. The birds and animals that had long disappeared from the program area have returned, and the wild animals have recovered greatly both in species and population.

巍巍太行·河北邢台
The majestic Taihang Mountains — Xingtai, Hebei

太行山是中国东部一条南北走向的重要自然地理分界。东面的华北平原是落叶阔叶林地带，西面的黄土高原是森林草原地带和半干旱草原地带。
The Taihang Mountains are the important geographic boundary that goes in north-south direction in the eastern China. The North China Plain in their east belongs to deciduous broad-leaved forest zone, and the Loess Plateau in their west belongs to forest/grassland and semi-arid grassland zones.

华北区 North China Region

燕山冬日·北京金山岭
The Yanshan Mountains in winter
Jinshanling, Beijing

延绵的燕山山脉以落叶阔叶林占主体地位，旭日升起，白色的雪岭染上了一片红褐色。
The endless Yanshan Mountains feature deciduous broad-leaved forest mainly. Rising sun tints the snow mountains into a scarlet cover.

秋色斑斓·河北隆化
Colorful autumn
Longhua, Hebei

暖温带的森林多由落叶阔叶林和针叶落叶阔叶混交林组成，秋天一到，漫山遍野尽显色彩斑斓。
Forests in warm temperate climate zone are composed of deciduous broad-leaved forests and mixed coniferous deciduous broad-leaved forests. When autumn comes, the mountains are covered with gorgeous multi-colored forests.

滩涂湿地·黄河三角洲
Tidal wetlands
Yellow River Delta

河口、滩涂等大量湿地是芦苇、碱蓬等湿地植物生长的优良环境，滋生了大量的鱼、虾、蟹、螺、贝等，为种群数量繁多的迁徙鸟类提供了充足的食物。
A large number of wetlands such as estuaries and tidal flats are good environment for wetland plants such as reeds and *Suaeda glauca*. They are home to vast amount of fish, shrimps, crabs, snails and shellfish, providing abundant food for various migratory birds.

华北区 北临蒙新区与东北区，南抵秦岭—淮河一线，西起西倾山，东至大海，包括了太行山、燕山、伏牛山、六盘山等山脉，以及黄土高原、晋冀山地、黄淮平原和黄海、渤海两海大部，属世界动物地理区划中古北界的中国东部分，气候带为中温带和暖温带。这里是中国南北动物区系的过渡带，西北部以中亚型占优，西部以高地型和喜马拉雅—横断山区型占优，南部以南中国型和东洋型占优，全区以华北型或季风区型占比最高。

该区的渤海湾、黄河三角洲、苏北滩涂的滨海迁徙鸟类极多，是东亚—澳大利西亚水鸟迁徙路线的重要组成部分，横贯该区的黄河沿岸湿地也是大量迁徙鸟如天鹅等途经的地方。这些湿地为数百种、千万只的迁徙水鸟提供了优良的中途停歇地，使其得以补充和储备继续飞行所需的能量。另外，渤海辽东湾、山东半岛北部海域是全球斑海豹8个繁殖区之一，也是其分布区的最南端，生活在这里的西太平洋斑海豹是唯一在中国海域进行繁殖的鳍足类动物。

但是，该区域人口密度较大，有数千年的农垦开发历史，致使农田生态系统非常发达且面积占比很大，野生动物的栖息地受人类干扰尤其严重，很多地方野生动物几乎消失殆尽。其占主体的黄河中下游流域（相比华中区的长江流域）光热水条件较差，生态环境破坏后恢复很难，因此，该地区野生动物物种较为稀少，尤其是平原地区已经很难寻觅到较大体型野生动物的踪迹了，相比起来，山区的野生动物状况会好一些。这些年来，由于对野生动物保护力度的不断加大，尤其是退耕还林还草及野生动物保护和自然保护区建设工程的实施，山区常常发现有关键种——华北豹的活动踪迹，显示这里的自然生态环境有了较大好转。2019年7月，黄（渤）海候鸟栖息地（第一期）列入世界遗产，成为我国首块、全球第二块潮间带湿地世界遗产。中国最后的麋鹿于20世纪初在北京南海子消失后，又于20世纪末在原地及江苏大丰重引入成功，目前江苏大丰具有世界上最大的麋鹿种群。

随着国家建设生态文明社会重大战略的逐步推进和具体措施的逐步落实，"绿水青山就是金山银山"的理念深入人心，"山水林田湖草沙"统一部署，给该地区带来了生态环境建设新的机遇和动力，野生动物在种类和数量上都逐步得到了恢复。

The North China Region (NCR) borders on the Inner Mongolia and Xinjiang Region (IMXR) and the North-east Region (NER) in the north, Qinling Mountains-Huaihe River in the south, the Xiqing Mountains in the west and the sea in the east. In addition to such mountains as the Taihang, the Yanshan, the Funiu and the Liupan, the NCR is also home to a highly diversified landscape that includes the Loess Plateau, the hilly areas in Hebei and Shanxi provinces, the Huanghuai Plain, and a major part of the Yellow and the Bohai seas as well. It belongs to the Palaearctic Realm of the eastern China in global zoogeographical zoning system, covering both mid-temperate and warm temperate climate zones. Situated on the borderlines between the southern and the northern fauna in China, this region can be further broken down into several subsections that have their dominating fauna species. Specifically speaking, whereas the northwestern subsection, the southwestern subsection and the western subsection of the NCR regions are respectively dominated by central-Asian species, highland and Himalaya-Hengduan species, and Southern China and oriental species, the defining species of the entire region fall into either North China species or monsoon species.

The Bohai Bay, Yellow River Delta and tidal flats in the northern Jiangsu within the NCR are places where vast number of migratory birds pass by each year, making this region a key section of the global East Asian-Australasian Flyway (EAAF). The wetlands along the Yellow River provide an excellent stopover for the tens of millions of migratory waterfowls such as swans, enabling them to replenish energy and store fat needed for their long-distance flight. In addition, among the eight breeding areas for spotted seals (*Phoca largha*) in the world, the one that lies between the Liaodong Bay and the northern sea of the Shandong Peninsula is the southernmost one. The spotted seal is the only pinniped mammal that can breed in seas within Chinese territory.

However, because of the relatively higher population density and long-rooted agricultural history in the NCR, human-inflicted disturbance to wildlife habitats is especially serious, leading in some places even to nearly extinction of wild animals. Compared with the Yangtze River Basin in the Central China Region (CCR), the middle and lower reaches of the Yellow River basin that makes up the major part of NCR has relatively poorer sunlight, heat and water resources, and is therefore more vulnerable in its ecological environment. As a result, large-sized wildlife species are now rarely seen in this region, especially in plain areas. The situation in mountainous areas is comparatively better. The past few years have witnessed increasing efforts in the protection of wild animals, which is particularly the case since the implementation of Conversion of Cropland to Forest/Grassland Program and Nature Reserve Development Program. For example, traces of North China leopard (*Panthera pardus fontanierii*), a keystone species, have often been found in mountainous areas lately, which shows that the natural ecological environment here has been greatly improved. In July 2019, the Migratory Bird Sanctuaries along the Coast of Yellow Sea-Bohai Gulf (Phase I) was listed as a world heritage site, the first among its counterparts in China, and the second intertidal wetland world heritage in the world as well. After the last Père David's Deer (*Elaphurus davidianus*) disappeared in Nanhaizi in Beijing in early 20th century, the Chinese government has re-introduced the species to Nanhaizi in Beijing and Dafeng in Jiangsu successfully. Nowadays, Dafeng has the world's largest population of the species.

Following the steady progresses made in China's key strategy of developing ecological civilization, and thanks to the concrete measures that have been taken in well-paced schedules, ideas like "lucid waters and lush mountains are invaluable assets" and "co-ordinated management of mountains, rivers, forests, farmlands, lakes, grasslands and deserts" have now become deeply rooted in people's mind. All these have brought about new opportunities and injected new impetus for the ecological improvement endeavors in the region, with notable growth obtained in both the diversity and population of wild animals.

↑ 野放的公麋鹿群·江苏大丰
Male Pere David's deer reintroduced to the wild — Dafeng, Jiangsu

在大丰这片广阔的滩涂湿地上，经人为引入已经适应野外生活的雄性麋鹿自由自在地集群、奔跑、觅食和繁衍着。这里生活着世界上最大的麋鹿种群。

In this vast tidal wetland, the male Pere David's deer (*Elaphurus davidianus*) reintroduced back into the wild have adapted themselves perfectly to their new environment, where they converge, run, forage and multiply freely. The largest Pere David's deer population is living there.

传奇的麋鹿故事
Legends of Pere David's Deer

鹿鹭对话·江苏大丰

Dialogue between the deer and the heron — Dafeng, Jiangsu

一只成年雄性麋鹿在和一只牛背鹭交流对话，它们对现在自由自在的生活应该还满意吧。

An adult male Pere David's deer is having a dialogue with a cattle egret (*Bubulcus ibis*), both of whom seem to be pretty satisfied with the carefree life they are living now.

麋鹿在中国文化中可谓源远流长，周朝灭纣时姜子牙的坐骑就是麋鹿。屈原有"麋何食兮庭中？蛟何为兮水裔？"之辞，全唐诗中关于麋鹿的诗句近百处。麋鹿之所以圈养于皇家猎苑，乃与皇权帝位及对于福禄的期盼有关，清乾隆皇帝更有"岁月与俱深，麋鹿相为友"的诗句。

麋鹿原产于中国，其经历充满了传奇色彩。据科学考证，早在3000多年前，我国黄河、长江中下游地区广有麋鹿，但汉朝以后逐渐减少。至清朝时期，为了供皇帝游猎，残余的麋鹿被捕捉到皇家猎苑内饲养，到19世纪时，就只剩北京南海子皇家猎苑内一群了。1866年，该群麋鹿被法国传教士大卫神甫在此"发现"并命名了拉丁种名，其影响震惊西方世界。之后，各国公使纷纷采用贿赂、偷盗等手段，先后将麋鹿带回自己国家以供猎奇观赏。直至1900年，"八国联军"入侵北京，最后一群麋鹿惨遭厄运，有的被杀戮，有的被装上西去的轮船。从此，麋鹿在中国完全绝迹。

后来，流落在国外的麋鹿大部分相继死去，只有英国贝福特公爵私人别墅的乌邦寺里饲养的麋鹿生长良好，达400多头，并向各国输出。1986年8月，在世界自然基金会和中国林业部（现国家林业和草原局）的努力下，我国从英国乌邦寺迎归了20头麋鹿，放养在清代曾豢养麋鹿的北京南海子。1987年8月，英国伦敦动物园又无偿提供了39头麋鹿，放养在江苏大丰麋鹿自然保护区。至此，麋鹿结束了它们大半个世纪在海外漂泊不定、颠沛流离的生涯，开始了回归故土、回归大自然的新生活。

中国麋鹿现主要分布在四个地方，即江苏大丰、北京南海子、湖北石首和湖南洞庭湖。其中，面积达7.8万公顷的江苏大丰麋鹿自然保护区，是世界上面积最大、麋鹿种群最大的地方。麋鹿回归的那一段时间，我曾经参与了其中部分工作。1998年，我主持并目睹了大丰的首次麋鹿野外放归活动，野外放归的成功标志着我国已经建立了可自我维持的麋鹿野生种群，结束了数百年来全球麋鹿无野生种群的历史。归去来兮，传奇麋鹿野外放归的成功是中国乃至世界野生动物重引入的成功案例！

The Pere David's deer, or Milu deer in Chinese, is a deep-rooted element in Chinese culture. It is believed that Jiang Ziya, a legendary figure in the Zhou Dynasty who put an end to the tyrannical rule of the Shang Dynasty, used to ride on this very animal as his mount. Quyuan, the great writer that lived during the Warring States Period, celebrated highly on this animal in his well-quoted ode, "For what reason Milu deer is fed in people's courtyard? For what reason dragons are regarded as the descendants of the sea?" In the *Complete Collection of Poems of the Tang Dynasty*, one will find nearly 100 poems in which the Milu deer was mentioned. Besides, Milu deer are often kept in royal hunting gardens and regarded as symbols of the royal power as well as people's expectation for happiness and prosperity. Emperor Qianlong of the Qing Dynasty wrote in one of his poems a line that goes as follows: as time moves on, bonds between me and Milu deer grow stronger.

Being the place where Milu deer has its origin, China is teeming with legendary stories related with the animal. Scientific investigations reveal that, as early as over 3,000 years ago, it was already widely found in the Yellow River region and the middle and lower reaches of the Yangtze River in China, though with gradually declining population since the Han Dynasty. During the Qing Dynasty, the small number of deer that still survived were captured and kept in royal hunting gardens for the emperors to game. By the 19th century, the deer in the Nanhaizi Royal Hunting Garden in southern Beijing had becoming the only group that was known to have survived. In 1866, Pere David, a French missionary, "discovered" the animal and gave it a Latin name. In the wake of this discovery that intrigued the attention of the West, envoys from some countries resorted to bribery, theft and other means to smuggle the deer out of China into their own countries as an exotic species. In 1900, when the Eight-Nation Alliance invaded Beijing, a catastrophic toll was done on this only surviving Milu deer community. With the animals either cruelly slaughtered or shipped westward to other countries, Milu deer went extinct in China.

Later on, most of the deer smuggled abroad died. The only group that thrived was the ones in the Woburn Abbey, private villa owned by the Duke of Bedford, whose population grew up to 400 and were exported to other countries. Under the joint efforts of the

WWF and the former Chinese Ministry of Forestry (the present National Forestry and Grassland Administration), 20 heads of Pere David's deer were imported in August 1986 from the Woburn Abbey and introduced to Nanhaizi in Beijing, which used to be the royal hunting garden where the deer were kept during the Qing Dynasty. In August 1987, London Zoo donated 39 heads of deer, which are kept in the Dafeng Pere David's Deer Nature Reserve in Jiangsu Province, marking an end to the half-a-century exile of Pere David's deer and ushering in a new chapter for the animal to live in the natural habits of their home country.

Pere David's deer in China are currently distributed in four key regions — Dafeng in Jiangsu Province, Nanhaizi in Beijing, Shishou in Hubei Province and Dongting Lake in Hunan Province. Dafeng Pere David's Deer Nature Reserve, covering an area of 78,000 ha, is the largest nature reserve in terms of both its area and deer population. I took part personally in the early stage of efforts to bring the deer back to China and hosted and witnessed the first endeavor in 1998 to reintroduce deer raised in captivity in Dafeng back into the natural environment. The success of this event marked the establishment of self-sustainable wild Pere David's deer population and put an end to the century-old absence of wild Pere David's deer population in the world. The returning of this legendary deer back to China — its home country, their successful reintroduction back to nature that followed shortly afterwards, all these are indeed cases that both China and the global community can draw on in their efforts in this field.

唯一的海豹
The Unique Seal

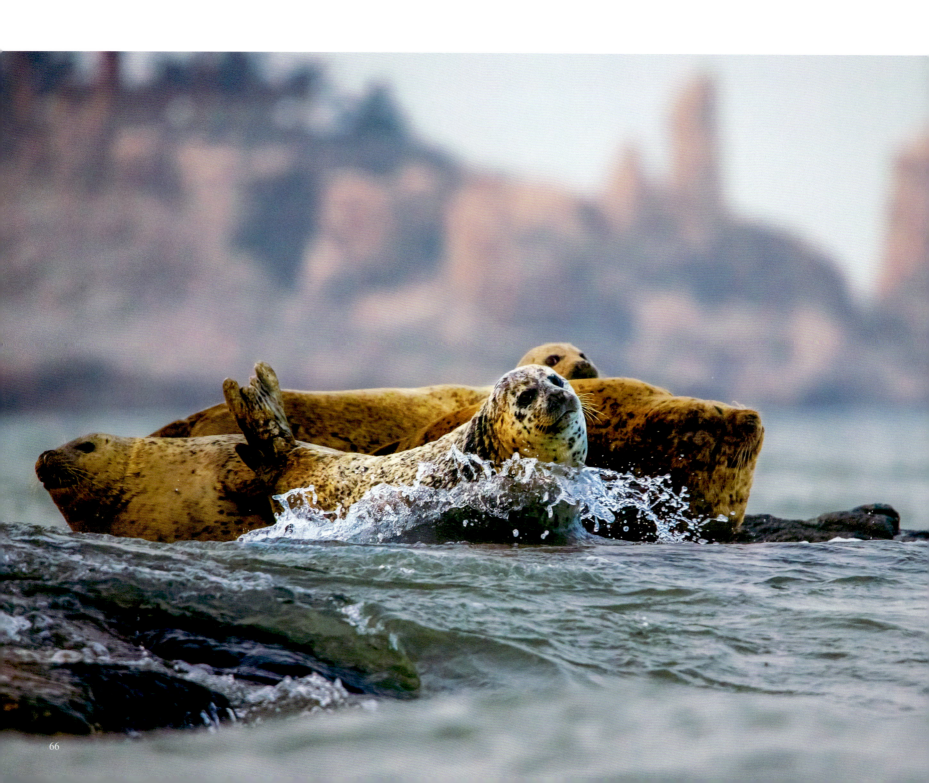

西太平洋斑海豹是唯一能在中国海域繁殖的鳍足类动物，是渤海、黄海海洋生态系统的指示种，尤为珍贵。它们主要生活在北半球的高纬度地区，我国的渤海辽东湾、山东半岛北部海域是全球斑海豹8个繁殖区中最南的一个。斑海豹有洄游的习性，它一生的大部分时间是在海水中度过的，仅在生殖、哺乳、休息和换毛时才爬到岸上或冰块上。西太平洋斑海豹主要以捕食鱼类为主，也食甲壳类及头足类生物。渤海海岸滩涂和庙岛群岛的岩石礁，以及冬季渤海湾里形成的浮冰块为西太平洋斑海豹提供了栖息繁殖的好场所。

本区西太平洋斑海豹的数量曾经十分庞大，但是由于受到长期过度捕捞，特别是对幼仔的猎捕，西太平洋斑海豹资源遭到了严重的破坏，其数量急剧下降。近些年来，工农业对于沿海滩涂的大肆围垦开发，河口、海洋的污染，航运量的迅速增加破坏了西太平洋斑海豹的栖息环境，以及无节制捕获海洋生物等原因致使西太平洋斑海豹食物减少，这些都对西太平洋斑海豹的生存构成了极大的威胁。

国家及地方政府采取了积极的应对措施，1982年成立了山东长岛自然保护区，总面积达5015公顷，其中，海域面积1105公顷，保护了部分西太平洋斑海豹栖息地。1997年，在渤海沿岸建立了中国唯一的以西太平洋斑海豹为主要保护对象的国家级自然保护区，范围涉及辽东半岛南端的海岸、海域以及70多个岛屿，总面积为90万公顷，有效地保护了栖息在该地区的西太平洋斑海豹。近些年来，主管部门及有关地方政府坚持保护与增殖放流措施相结合，打击非法猎捕活动，控制环境污染，使西太平洋斑海豹的濒危状况得到了较大缓解。

The spotted seals are particularly precious since they are the only pinniped mammal that can breed in seas within Chinese territory as the ecological indicator species of the ecosystem in the Yellow Sea and the Bohai Sea. They mainly live at high latitudes of the northern hemisphere. Among the eight breeding areas for spotted seals in the world, the one that lies between the Liaodong Bay and the northern sea of the Shandong Peninsula is the southernmost one. They have the habit of migration, spending most of their lifetime in the sea, and coming ashore or climbing up to ice blocks only during the lactation periods, or when they need

a rest or are in molt. Fish, crustaceans and cephalopods are their food. The tidal flat of the Bohai Sea, the shore reef of the Miaodao Islands and floating ice-cubes in the Bohai Gulf during winter times provide good places for spotted seals to inhabit and breed.

The population of spotted seals inhabiting the NCR region used to be very large. But the past few decades have witnessed a drastic drop due to chronic overfishing, especially that of the young seals. In recent years, the spotted seals are faced with tremendous risks that endanger their survival as the tidal flats are developed excessively to meet industrial and agricultural needs, their habitats are increasingly damaged by pollutants discharged at river estuaries and oceans as well as by ever busier shipping activities, and food supply for the seals are getting increasingly scarce due to the exploitative utilization of marine resources by humans.

In response, the national and local governments have taken active measures. The Changdao Nature Reserve of Shandong Province was set up for the habitat protection of partial spotted seals in 1982, covering a total area of 5,015 ha, including 1,105 ha of sea area. In 1997, China set up its first and only national nature reserve (NNR) that takes spotted seals as the primary target of protection. This seal-featuring NRR, 900,000 ha in size, covers the sea, coasts and over 70 islands that lie to the south of the Liaodong Peninsula in the Bohai Sea, providing a safe haven for the spotted seals inhabited in this area. In recent years, authorities and local governments have adopted a series of measures that focus both on protective efforts like cracking down on illegal poaching, reducing pollutant discharge and *etc.*, and on population-boosting efforts like introducing seals bred in captivity into nature, all of which have greatly mitigated the endangered status of the spotted seals.

← 迎新·山东长岛
Greeting the rising sun — Changdao County, Shandong

日出东方，礁石上的西太平洋斑海豹又迎来了新的温暖的一天，惬意的眼神和动作充满了对于现今生活的满足感。

As the sun rises in the east, the spotted seal (*Phoca largha*) on the reef ushers in another warm day. A sense of content can obviously be felt in its eyes and through its movement.

← 家园·山东长岛
Homeland — Changdao County, Shandong

渤海湾是西太平洋斑海豹繁殖的美好家园，冬季就要过去，产仔期就要到来，西太平洋斑海豹群在岛礁上享受着海浪的拍打和春天温暖的阳光。

The Bohai Bay is a wonderful breeding home for spotted seals (*Phoca largha*). As cold winter days are about to give way to spring — their breeding season, the seals on the reefs are taking their time to relax in the gentle waves and balmy sunlight.

种类数量繁多的滨海迁徙鸟类
Rich Variety of Coastal Migratory Birds

飞行马拉松冠军·河北北戴河
Champion of marathon flight — Beidaihe, Hebei

斑尾塍鹬可以从新西兰途经渤海湾歇息后飞阿拉斯加，再由阿拉斯加直飞新西兰，连续不停地飞行11500多千米，这是鸟类不间断飞行的世界最长纪录。

The bar-tailed godwit (*Limosa lapponica*) is capable of flying from New Zealand via the Bohai Bay to Alaska, from where they will fly directly back to New Zealand. A non-stop flight that covers more than 11,500 km in distance! This is the longest distance by birds that has ever been recorded in the world.

黄渤海湿地是中国渤海湾及黄海滩涂湿地的总称，是全球最重要的沿海湿地生态系统之一。黄渤海湿地既是水鸟沿海岸线迁飞至长江中下游地区或华南越冬的停歇地，也是水鸟继续南飞至东南亚、澳大利亚和新西兰等地越冬的停歇地，是东亚—澳大利西亚水鸟迁徙路线的最重要组成部分。从世界重要候鸟迁徙路线来看，在俄罗斯、日本、朝鲜半岛和我国东北与华北东部繁殖的湿地水鸟，春秋两季主要通过我国这个地区进行南北方向的迁徙。在春季，来自南太平洋诸岛和大洋洲的北迁鸟类到达中国台湾后，分为两支，主要一支沿中国大陆扩散或继续沿东部海岸北上，另一支经琉球群岛到日本或继续北迁。沿中国大陆东部沿海北迁的鸻鹬类等湿地水鸟在到达长江口以后，又分两条北上路线迁徙：主要一条经江苏、山东、河北沿海到我国东北地区和俄罗斯，另一条则越海向朝鲜半岛或日本迁飞。在秋季，湿地水鸟沿中国东部沿海向南迁飞至华东和华南，远至东南亚各国，或由俄罗斯东部途经中国向东南亚至澳大利亚迁徙，其南下迁徙路线大致与春季北上路线相似。

黄渤海湿地广阔的海岸滩涂为数百种数千万只的迁徙水鸟提供了优良的中途停歇地，使其得以补充能量、储备继续飞行所需的脂肪。仅在华北区沿海停歇的鸟类中，鸻鹬类的鸟就有25种和超过全球总数30%的种群数量，有大约80%的大杓鹬和40%的半蹼鹬在此地停歇，而斑尾塍鹬、弯嘴滨鹬、白腰杓鹬、大滨鹬和环颈鸻的数量甚至是其迁徙种群的全部。这些滨海迁徙鸟类中，有各种各样的传奇故事，例如，斑尾塍鹬就是鸟类的飞行马拉松冠军，创造了鸟类不间断飞行的世界最长纪录。

2019年，黄（渤）海候鸟栖息地（第一期）被列为"世界遗产"，这是中国首块、全球第二块潮间带湿地"世界遗产"。

Wetlands of the Yellow and Bohai Seas (WYBS), which is a general term that refers to all the tidal wetlands lying in the Bohai Gulf and along the Yellow Sea, is one of the world's most important coastal wetland ecosystems. As a key section of the global East Asian-Australasian Flyway (EAAF), WYBS is not only an important stop for waterfowls that migrate along the coastlines to the middle and lower reaches of the Yangtze River and other parts of the South China Region for overwintering, but also an important stop for those that migrate further southward to overwinter in Southeast Asia, Australia and New Zealand. From the perspective of the important migratory routes around the world, waterfowls that breed in wetlands in Russia, Japan, the Korean Peninsula, and the North-east Region (NER) as well as the eastern part of the North China Region (NCR) in China stop over the WYBS on their southward or northward migratory journeys during the spring and autumn seasons. Upon their arrival at Chinese Taiwan in spring, the northbound migratory birds from islands in the South Pacific and Oceania will divide into two groups: one group will either spread out across the Chinese mainland or fly further northward along the coastlines of eastern China; the other group will either fly to Japan via the Ryukyu Islands or continue further on their northbound migration. The group that fly northward along the coastlines of eastern China, typically composed of fowls that fall into the family of waders, will split once again upon arriving at the Yangtze River estuary, with some flying further northward to the northern China and Russia via the coastal provinces like Jiangsu, Shandong and Hebei, and the rest flying to the Korean Peninsula or to Japan across the seas. In autumn seasons, waterfowls will fly southward along Chinese coastlines to the East China Region (ECR) and the South China Region (SCR) in the country, or further to other countries in Southeast Asia. At the same time, birds that

滩涂鸟浪·江苏盐城
Rolling sea of waterfowls on tidal flats — Yancheng, Jiangsu

在海河共同作用下形成的大片滩涂上，咸水、半咸水的环境中丰富的营养物质养育了无数的小生物。这里是大批迁徙鸟类最好的营养加油站，成千上万的水鸟蜂拥而上，构成了一阵阵翻腾的鸟浪，遮天蔽日，蔚为壮观。

Thanks to the joint shaping of both the river and the sea, vast stretches of tidal flat come into being. Endowed with a nutrient-rich environment that has given birth to vast number of creatures that live in salty and semi-salty water, tidal flats here make up an ideal top-up stop for migratory birds that pass by. Thousands upon thousands of waterfowls gather on the flats to form rolling sea of birds that extends until the end of horizon. What a magnificent scene it is!

started their southbound migration from the eastern part of Russia to Southeast Asia and Australia will also pass by the WYBS in China, with a migratory route roughly similar to their northward route during spring seasons.

Vast stretches of tidal flats in the WYBS area provide an excellent stopover for the hundreds of millions of migratory waterfowls, enabling them to replenish energy and store fat needed for their long-distance flight. Take the waterfowls that stop by in the NCR for example, they make up 25 waders species, about 30 percent of the world's total wader population; about 80% of eastern curlews (*Numenius madagascariensis*); and 40% of Asian dowitchers (*Limnodromus semipalmatus*). As to the bar-tailed godwit (*Limosa lapponica*), curlew sandpiper (*Calidris ferruginea*), Eurasian curlew (*Numenius arquata*), great knot (*Calidris tenuirostris*) and Kentish plover (*Charadrius alexandrinu*), almost all the migratory birds under these species will stop by this region. Each of these coastal migratory birds has its legendary stories. The bar-tailed godwit, for example, is known as marathon champion among all birds, which set the record in the distance of non-stopping flights in the world.

The Migratory Bird Sanctuaries along the Coast of Yellow Sea-Bohai Gulf (Phase I) was listed as a world heritage site in 2019, the first among its counterparts in China, and the second intertidal wetland world heritage in the world as well.

飞行的白琵鹭群·黄河三角洲
A flock of white spoonbills in the air — Yellow River Delta

白琵鹭为大型涉禽，为夏候鸟，繁殖于中国北方包括西藏等地，越冬于中国南方的华中、华南区。因其扁平的长嘴与中国乐器中的琵琶极为相似而得名。

The white spoonbill (*Platalea leucorodia*), a large-sized wader, is a summer resident, which breeds in the northern China including Tibet and winters in the southern China including the Central China Region and the South China Region. It is named in Chinese as Bai Pi Lu because the long and flat beak that the bird has looks very much similar to the shape of a traditional Chinese musical instrument — pi pa.

多种多样的湿地鸟类·黄渤海湿地
A big variety of wetland birds — Wetlands of the Yellow and Bohai Seas

这里有全世界规模最大的潮间带滩涂，是2000多种动物的栖息地，也是西伯利亚、东亚与澳大利亚一带候鸟迁徙最重要的中转站，数千万候鸟每年在这里停歇、换羽或度过寒冷的冬季，如：白腰杓鹬（图1）、蛎鹬（图2）、青脚鹬（图3）、普通海鸥（图4）、青头潜鸭（图5，拍摄地为河北衡水湖，全球极度濒危物种，数量不足千只）、黑嘴鸥（图6，拍摄地为黄河三角洲，濒危物种，数量稀少，繁殖地主要在中国）、反嘴鹬（图7）、大麻鳽（图8）、林鹬（图9）、凤头麦鸡（图10）、金眶鸻（图11）等。

Being the world's largest intertidal flat, this is a most-favored habitat for over 2,000 species of wild animals as well a critical stop for the migratory birds that shuttle annually between Siberia and Australia via the eastern Asia. Millions of migratory birds rest, moult, or spend the cold winters here each year. Shown in photos on the right page are just a few of the examples: Eurasian curlew (*Numenius arquata*, Photo 1), oyster catcher (*Haematopus ostralegus*, Photo 2), common greenshank (*Tringa nebularia*, Photo 3), common gull (*Larus canus*, Photo 4), Baer's pochard (*Aythya baeri*, Photo 5, taken at Hengshui Lake in Hebei. This is a critically endangered species of the world whose population adds up to no more than 1,000.), Saunder's gull (*saundersilarus saundersi*, Photo 6, taken at the Yellow River Delta. This is a rare and endangered species that takes China as its primary breeding place.), pied avocet (*Recurvirostra avosetta*, Photo 7), common bittern (*Botaurus stellaris*, Photo 8), wood sandpiper (*Tringa glareola*, Photo 9), northern lapwing (*Vanellus vanellus*, Photo 10), and little ringed plover (*Charadrius dubius*, Photo 11).

← **黑尾鸥的自由王国·长岛鸟岛**
Kingdom of the black-tailed gulls — bird islands, Changdao County

在渤海的诸多岛屿中，不仅有蛇岛，更有很多鸟岛。鸟岛往往由一两种鸟作为优势种，其数量巨大，飞起来遮天蔽日。这是长岛群岛中一座以黑尾鸥为主的鸟岛，这里是它们的繁殖地和自由王国。

The Bohai Sea is not only home to the snake island, but also home to numerous bird islands. The bird islands here often has one or two dominant bird species that are extremely large in populations. Shown in this photo is one of the bird islands among the Changdao Archipelago, which is dominated by and provides a free kingdom for the black-tailed gulls (*Larus crassirostris*).

飞翔的大鸨群·河南郑州
Great bustards on the wing — Zhengzhou, He'nan

大鸨为大型地栖鸟类，主要栖息于干旱草原、稀树草原和半荒漠地区，但在华北平原的农田中也能够看到它们的踪影。

The great bustard (*Otis tarda*), a large terrestrial bird, mainly lives in arid grasslands, savannas and semi-desert areas. But it is also often found in farmlands of the North China Plain.

等待中的杀机·山东长岛
Looming danger — Changdao County, Shandong

长岛的特有蛇类——庙岛蝮属树栖性蛇。候鸟迁徙季节，蝮蛇会在树枝上静静地等待，一旦迁徙的小鸟落下，它就会以迅雷不及掩耳之势扑向猎物。

Gloydius lijianlii is an endemic arboreal snake typically found in Changdao County. During migration seasons, it will wait patiently in trees for chances to prey on birds that fall unwittingly into its traps.

鸿雁飞来·河南黄河湿地
The arrival of swan geese from afar — wetlands of the Yellow River, He'nan

"鸿雁向南方，飞过芦苇荡"，每年9月下旬至10月末，鸿雁就开始大量地从繁殖地南迁往黄河周边的芦苇荡越冬。

"South-bound swan geese flying over endless reed marshes" is a familiar scene that takes place each year in late September until the end of October in the northern China, when thousands upon thousands of swan geese (*Anser cygnoid*) would flock together from their breeding land southward to overwinter in the reed marshes around the Yellow River.

水上激情·北京野鸭湖
Passion on the water — Yeya Lake, Beijing

凤头䴙䴘因有显著的黑色羽冠而与普通䴙䴘有别。繁殖期，它们成对地作精湛的求偶炫耀，身体高高挺起，在水面上划动舞蹈，两相对视并不断点头，激情高潮时交配。

The great crested grebe (*Podiceps cristatus*) differs from the common grebes in that they have distinctive black crowns over their heads. During mating seasons, couples of this species will often present highly romantic shows of courtship, in which the two fowls will dance elegantly over the water, looking at each other and nodding their heads, until they both reach the right emotion for mating.

第三者·江西鄱阳湖
"The other man" — Poyang Lake, Jiangxi

在自然界，大白鹭的花心偶见，红杏出墙也是有的。请你判断一下，谁是第三者？

The great egret (*Ardea alba*) might occasionally betray its loved ones and have affairs with someone that intrudes into the relationship. Can you see which among them is "the other man"?

幸福夫妻·河北衡水湖
A lovely couple — Hengshui Lake, Hebei

灰翅浮鸥的巢通常为飘浮于水中植物上的浮巢，巢材为芦苇、蒲草等水生植物。图为幸福的灰翅浮鸥夫妇守护在爱情的结晶旁。

The *Chlidonias hybrida* tends to build their nests, typically made from aquatic plants like reeds and narrowleaf cattails (*Typha angustifolia*), on plants floating in the water. The couple in the photo are looking tenderly at their egg and guarding the crystallization of their love.

↑ 华北的雉类·华北区
Pheasants in the northern China — NCR

中国是雉类王国，尤以西南区为胜，但有些雉类仅分布于华北区，如褐马鸡（图2，拍摄地为山西芦芽山）仅见于山西、陕西及附近地区；白冠长尾雉（图1，拍摄地为河南连康山）主要分布在河南董寨、连康山等；红腹锦鸡（图3，拍摄地为陕西宝鸡）主要分布在华北；只有环颈雉是广布物种，但在华北区尤盛。

China is a favored kingdom for pheasants. Though the region with the greatest diversity of pheasants in the country is the SWR, some pheasant species are only distributed in the NCR, as shown in the photos on this page: the brown-eared pheasant (*Crossoptilon mantchuricum*, Photo 2, taken at the Luya Mountains in Shanxi), which is found only in Shanxi, Shaanxi and surroundings areas; the Reeve's pheasant (*Syrmaticus reevesii*, Photo 1, taken at Liankang Mountains in He'nan), which is mainly distributed at Dongzhai and Liankang Mountains in He'nan; and the golden pheasant (*Chrysolophus pictus*, Photo 3, taken at Baoji in Shaanxi), which lives mainly in the NCR. The ring-necked pheasants (*Phasianus colchicus*) is the only species that is extensively distributed throughout the country, though primarily in the NCR.

新添的北京户口·北京奥林匹克森林公园 →
Newly-arrived dwellers in Beijing — Beijing Olympic Forest Park

北京奥林匹克森林公园是为2008年奥运会修建的人工园林公园，随着公园生境的逐渐优化和近自然化，野生动物也多了起来，不知不觉之间已经有不少环颈雉在此谈情说爱、生儿育女。

The Beijing Olympic Forest Park is an artificial forest park built specially for the 2008 Olympic Games. As the environment in the park is gradually improved and optimized to a near-nature state, ring-necked pheasants (*Phasianus colchicus*) have increasingly fell in love and bred their children there.

| 生态思考 | 北京城的水鸟故事
ECOLOGICAL Reflection | The Story of Waterfowls in Beijing

1980年12月21日，在《北京晚报》报道了大天鹅首次来京并在玉渊潭过冬喜讯后的第四天，就传来了其中一只大天鹅被枪杀的噩耗。另外一只失去伴侣的大天鹅，嗷嗷不绝地哀鸣了整整一夜，在伴侣逝去的那块湖面上游来游去。第二天，这只天鹅翩然飞走，远离了这片伤心之地。这一事件随即引发了一场北京市民广泛自觉参与的野生动物保护的热议和行动。

如今的北京颐和园，来了多批次的大天鹅在昆明湖中停歇、飞翔，市民及游人们静静地欣赏着、欢迎着这些尊贵客人的到来，毫无惊扰之意。在"皇家花园落天鹅"这张照片里，有的大天鹅在昆明湖中游弋，更多的是把头插进翅膀里静静地休息，还有的正在空中准备落下。远方到来的大天鹅们把这里当成了自己的家，休息、觅食、补充营养、增加体力，呈现的是一片人与自然和谐的安详气氛。另外，还有诸如大白鹭自由飞翔在北京鸟巢前、鸳鸯安然生活在玉渊潭公园中等，这些是大中型水鸟在北京五环内的情况。五环以外的郊区，这些年途经的候鸟种群有2.5万~3万只，最高峰时一度达到5万多只，包括疣鼻天鹅、大小天鹅和赤麻鸭、豆雁等雁鸭类，仅鹤类就有灰鹤、白枕鹤和白头鹤，偶尔还能监测到白鹤，等等。这些事实生动地说明，随着中国社会文明的进步，北京市民对于野生动物保护的意识大大提高，使这个有2000万常住人口的国际大都市处处都呈现出人与野生动物自然和谐相处的崭新面貌。

On December 21, 1980, just four days after a report appeared in *Beijing Evening News* announcing the first-time arrival of whooper swans (*Cygnus cygnus*) in Yuyuantan Park for overwintering, one of the swans was reported to have been shot dead. Following the tragic death of her mate, the other heart-broken swan wailed bitterly throughout the night in the lake and flew away the next day. This sad event triggered a heated debate among residents in Beijing and kicked off an extensive campaign for wildlife protection.

Now in the Summer Palace in Beijing, there are many batches of whooper swans resting and flying around the Kunming Lake. Citizens and visitors silently appreciate and welcome the arrival of these distinguished guests without interruption. Swans in the Royal Garden depicts a touching scene in the Lake, with several whooper swans either swimming at leisure or resting in peace over the water, and some more in the air about to splash down. The whooper swans coming from afar have taken this place as their home, where they can rest and forage in carefree manners before setting out on their migratory journeys. What a peaceful and harmonious scene between human and nature it is! Photos telling about life of large and medium-sized waterfowls within the 5th Ring Road of Beijing can also be found in the album, such as the one showing a great egret soaring over the National Stadium, the one in which mandarin ducks swim elegantly over the water in the Yuyuantan Park. Beyond the downtown area, 25,000 – 30,000 birds stop over in the suburbs of the city on their annual migratory journeys. At peak times, birds passing by even amount up to more than 50,000, among which are anatidae such as mute, whooper and tundra swans, and ruddy shelduck, bean geese, as well as Gruidae such as common, white-naped and hooded cranes and occasionally even Siberian cranes. All these scenes testify strongly to the fact that, as civilization advances in the Chinese society, awareness of residents in Beijing for wildlife protection has been significantly improved. A brand new chapter marked by harmonious existence of human and wildlife has been opened in Beijing — an international metropolis with over 20 million permanent residents.

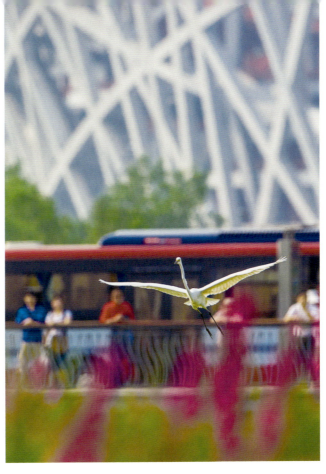

新公园新客人·北京奥林匹克公园
New guests to the new park — Beijing Olympic Park

北京奥林匹克公园是一个人工的体育公园，新公园迎来了新客人，小小大白鹭在大大"鸟巢"前自由地飞翔，吸引了众多路人驻足。

The newly built sports-themed Beijing Olympic Park in Beijing has attracted flocks of new guests. Against the giant National Stadium (nicknamed the Bird Nest), the great egrets (*Ardea alba*) soaring in the sky have become a new must-see view to visitors to the Park.

皇家花园落天鹅·北京颐和园
Swans landing on the royal garden — Summer Palace, Beijing

北京颐和园这座皇家园林，每天游人如织，但丝毫也没有影响来自远方的大天鹅飞来并在此休息、补充营养，毫无顾忌、自由自在地生活。

Whooper swans that have come from afar and settled down in the former royal garden — Summer Palace — swim at leisure in the lake, heedless of the visitors that flood daily into the garden for sightseeing.

复合生态系统·新疆哈密
The compound ecosystem—
Hami, Xinjiang

雪山、森林、沙漠、草原、湿地和野生动物六位一体的生态景观，是该区典型生态系统的集中统一代表，走遍中国甚至全世界也难得一见。
The six-in-one ecological landscape of snow mountains, forests, deserts, grasslands, wetlands and wild animals is the representative of the typical ecosystems in this area, which is rare in China and even in the world.

蒙新区
Inner Mongolia and Xinjiang Region

草原河曲·新疆巴音布鲁克
The meandering prairie river — Bayanbulak, Xinjiang

草原生态系统是"人—畜（包括野生动物）—草"三者共同协同进化的结果，自然、历史、科学和文化都告诉我们这一点。
Nature, history, science, and culture tell us that the grassland ecosystem is the result of the concerted evolution of "man, animal (including wildlife) and grass".

森林草原·新疆天山
The forest steppe — Tianshan Mountains, Xinjiang

森林草原是本区极具特色的生态景观，春天的森林草原在蓝天白云映衬下，山花烂漫格外美丽。
The forest steppe is especially distinctive ecological landscape of this area, especially under the blue sky and white clouds in spring, where the mountain flowers are luxuriantly blooming.

疏林沙地·内蒙古浑善达克沙地
The open forest sandy land — Hunshandake Sandy Land, Inner Mongolia

本区东部分布有很多沙地，榆树疏林和沙丘浑然一体是这里的典型生态景观。
There are a lot of sandy lands in the eastern part of this area, where the open elm forest and sand dunes are integrated, which is a typical ecological landscape of this region.

蒙新区 含内蒙古高原、鄂尔多斯高原、阿拉善高原、河西走廊、塔里木盆地、准噶尔盆地及柴达木盆地等，包括阴山、贺兰山、祁连山北坡、天山山地和阿勒泰山地等，属世界动物地理区划中古北界的中国北部分。该区气候带为中温带和暖温带，属半干旱、干旱和极干旱地区，自然条件相当恶劣，是我国最干旱且少雨少雪的地区。蒙新区的动物区系成分主要为中亚型、古北型和全北型，以耐干旱的动物为主，生存着一些干旱、半干旱地区所特有的荒漠、山地野生动物，有大型的有蹄类动物，如野骆驼、普氏野马、盘羊、蒙古野驴和天山马鹿等，也有众多的过境鸟和留鸟，如遗鸥、大天鹅、疣鼻天鹅等水鸟和鸶、隼、鹰、雕等猛禽。该区土地沙化及草原荒漠化比较严重，湿地萎缩退化、动植物种消失的总体趋势正在得到遏制，荒漠化和沙化面积连续15年实现"双缩减"。但是，部分地方水资源逐年减少、河流断流、湿地消亡、草原退化沙化，这严重地制约着野生动物的生存空间，更制约着当地社会经济的和谐发展。

因此，该区要在近年来取得防治荒漠化和自然保护成绩的基础上加大保护工作力度，三北防护林、天然林资源保护、京津风沙源治理、野生动植物保护、自然保护地建设和退耕还林还草生态工程更要继续精准、科学发力，植被恢复和建设中要着力实施好"宜林则林、宜草则草、宜湿则湿、宜荒则荒，大力防治荒漠化"的方针，在保护好现有植被的基础上，努力扩大林草覆盖度，从根本上扭转土地沙化和草原荒漠化扩展的趋势，努力在保持水土，保护好森林、湿地和草原生态系统，维护占比最大的荒漠生态系统的平衡，以及在保护生物多样性方面下狠功夫。

2018年10月，作为中国首批设立的10个国家公园体制试点之一——祁连山国家公园管理局揭牌，预示着该区自然生态和野生动植物保护工作拉开了崭新的一幕。今后，建议将天山、阿勒泰山、帕米尔、罗布泊国家公园等的建设列入议事日程，还要建设若干草原生态系统类型的国家公园以填补保护空白。中央"以国家公园为主体的自然保护地体系"建设战略在该区的具体落实并积极推进，必将使蒙新区的野生动物保护迈上一个新的台阶，迎来一个更加美好的明天。

Composed of the Inner Mongolia Plateau, Ordos Plateau, Alxa Plateau, Hexi Corridor, Tarim Basin, Junggar Basin and Qaidam Basin, the Inner Mongolia and Xinjiang Region (IMXR) is a region where numerous important mountains concentrate — Yinshan Mountain, Helan Mountain, the north slope of Qilian Mountain, Tianshan Mountain and Altai Mountain, to name just a few. According to the global zoogeographic zoning system, The IMXR belongs to the northern part of the Palaearctic Realm within China. Situated in the middle temperate zone and the warm temperate zone where the lands are typically semi-arid, arid and extremely arid, the natural conditions here are quite harsh, making it the driest and most rain-scarce place among all its counterparts in China. In terms of fauna system, the dominating wildlife species in the IMXR include those belonging to the Central Asia type, the Palaearctic type and the Holo-boreal type that are typically drought-tolerant and endemic to arid and semi-arid desert or mountainous areas, including large ungulates like wild camels (*Camelus ferus*), Przewalski's horses (*Equus ferus*), Argali sheep (*Ovis ammon*), Mongolian wild asses (*Equus hemionus*) and Tianshan red deer (*Cervus elaphus*), as well as a large number of migratory birds and resident birds, for example waterfowls like relict gulls (*Ichthyaetus relictus*), whooper swans (*Cygnus cygnus*) and mute swans (*Cygnus olor*), and predatory birds like buzzards, falcons, eagles and vultures, *etc*. Desertification of lands and grasslands, shrinking and degrading wetlands, loss of wildlife species are all on improving trends. Desertification of lands and grass lands has been on a declining trend for continuous 15 years. However in some places, water resources are decreasing year by year, rivers are drying out, wetlands are being lost, and grasslands are increasingly degrading into deserts, which not only imposes undesirable impacts on the survival of wildlife, but also gravely restricts the social and economic development of this region.

Therefore, efforts on nature protection should be further strengthened in the IMXR building on the achievements that have already been made over recent years in desertification combat and nature protection. Major ecological programs — like the Three-North Shelterbelts Program, the Natural Forest Protection Program, the Beijing-Tianjin Sandstorm Source Control Program, the Wild Fauna and Flora Protection Program, the Natural Reserve Development Program and the Conversion of Cropland into Forest/Grassland Program — must continue to play their respective roles on more scientific and targeted basis. In the process of vegetation restoration and rehabilitation, efforts should be made to put into practice the guiding principle for desertification-combating that advocates "establishing forests, grasslands, wetlands and maintaining deserts wherever it is suitable to do so". In addition to making sure existing grasslands are soundly protected, more efforts are needed to improve the vegetation coverage of this region, so that the trend of land degradation and desertification can be fundamentally reversed. The strictest measures must be taken to safely protect the ecosystems of forests, grasslands and wetlands, to maintain ecological balance of the desert ecosystem that takes up the largest proportion of lands within this region and to conserve biodiversity.

In October 2018, Qilian Mountain National Park Administration, which is among China's ten pilot national parks, was unveiled, marking the beginning of a brand new chapter in IMXR's work in ecological restoration and wildlife protection. It is recommended that the development of national parks in the Tianshan Mountain, Altai Mountain, Pamir and Lop Nur should be put on the agenda as soon as possible, and that the development of several national parks featuring the grassland ecosystem should also be considered to fill the conservation gap in this aspect. Following the step by step and proactive implementation of the central government's strategy for "natural protected area system highlighting the central role that national parks play", a brighter future can certainly be expected for wildlife protection in the IMXR.

↑ 儿马争雄·新疆卡拉麦里
Two horses competing for supremacy — Karamaili, Xinjiang

放归野外的普氏野马群在北疆卡拉麦里的雪野上重新开始生息繁衍。图为两匹儿马在雪原上争雄。
In the snow field of Karamaili, the northern Xinjiang, a herd of Przewalski's horses (*Equus ferus*) that had been reintroduced back to the wild started their new lives. The photo depicts two male horses competing against each other for supremacy in the snow field.

普氏野马重引入的成功
Successful Reintroduction of Przewalski's Horse

野马奔驰·新疆北塔山
Galloping wild horses — Beita Mountains, Xinjiang

放归野外的普氏野马群在荒漠雪原上自由自在地奔驰着。
The herd of Przewalski's horses reintroduced to the wild are galloping freely in the snowy desert.

2018年12月，困扰自然保护多年的石油开采彻底退出了新疆卡拉麦里自然保护区，这是政府牺牲了经济发展而作出的重大决策。这一决策使栖息在这里的野生动物包括普氏野马拥有了更加宽阔、自由的家园。

说起普氏野马回归自然的工作，我们走的是一条漫长而艰难的不凡之路，是多少自然保护人几十年默默奉献的结果，非常不易。普氏野马是世界上仅存的野马，具有6000万年的进化史，原产于中国新疆准噶尔盆地东部一带和蒙古国西部科布多盆地，被誉为"荒漠活化石"。20世纪60~70年代，蒙古国与中国先后宣布了本国普氏野马的灭绝。

1985年，为了让普氏野马重返家乡、重归自然，中国政府从欧洲引回了18匹人工环境下生活的野马，在新疆、甘肃两地进行半散放，为普氏野马重返大自然进行科学实验和先期研究，从而开始走上了野马驯养、繁育、野放的漫长道路。为改善普氏野马的基因状况，维护其遗传多样性，2005年，我国又从德国引入6匹人工环境下生活的雄性野马。我曾经负责分管过此事较长时间，年复一年，春夏秋冬，甚至在零下30摄氏度的极端天气下考察普氏野马的实际生存状况。而新疆野马繁殖研究中心的工作人员们为了普氏野马的野放工作，长年累月生活在茫茫戈壁里一个荒无人烟的地方，生活条件艰苦不说，由于平时生活单调，与人交流机会少，语言能力都下降甚至退化了，倒是与野马交流的手势、口令丰富多彩。我总是被他们所代表的无数中国自然保护基层工作者长年累月的牺牲和无私奉献精神所深深感动。

值得骄傲的是，他们30多年一贯坚持不懈的辛勤努力终于结出了丰硕的果实。截至2019年年底，我国普氏野马的野放种群已经不断扩大，成功繁育出6代714匹普氏野马，数量位居世界第一，并且已经有19匹普氏野马完全脱离人为干预，靠自己的能力在大自然中生存下来并且繁育了后代。普氏野马重引入并野放成功，是中国野生动物保护工作中值得称颂的重大成就。

In December 2018, oil exploration that had plagued nature conservation long was completely withdrawn from the Karamaili Nature Reserve in Xinjiang. This is a major decision made by the government at huge costs to its economic development. The decision gave a wider, freer home to wild animals including Przewalski's horse (*Equus ferus*).

The returning of Przewalski's horses to the wild has not been easy in coming by. It's the result of decades of dedicated work by many conservationists. As the only surviving wild horse in the world, Przewalski's horse has been going through a 60-million-year history of evolution. Originated in eastern Junggar Basin in Xinjiang of China and the Khovd Basin in western Mongolia, it is reputed as the "living fossil of deserts". In the 1960s – 1970s, Mongolia and China successively announced that Przewalski's horse had gone extinct in their countries.

In 1985, in order to bring Przewalski's horses back to their originating hometown and reintroduce them to the wild, the Chinese government bought 18 artificially-bred Przewalski's horses from Europe, which were then kept in semi-natural environments in Xinjiang and Gansu under a pilot project that aimed to reintroduce Przewalski's horses to natural environments and marked the beginning of the long and painful process for domesticating, breeding and reintroduction of Przewalski's horses. In 2005, China introduced six male Przewalski's horses from Germany to improve their genetic status and maintain their genetic diversity. I was for a long time responsible for this project, during which I had frequently been to the projects sites in all seasons year by year, even against the extreme weather when the temperature dropped to −30℃, to personally inspect the living environment of Przewalski's horses. As to the staff of the Xinjiang Wild Horse Breeding and Research Center posted year in and year out in the desolate places in the vast gobi, their lives were much tougher. In addition to coping with the harsh living conditions there, they also had to put up with the long and monotonous days when there were not even people around to talk with, resulting in notable degradations in their capacities for using human language but increasingly adept skills in communicating with Przewalski's horses through versatile gestures and sounds. I have always been deeply touched by the sacrifice and selfless dedication of countless grass-roots workers engaged in China's conservation endeavors, as showcased by them.

Much to our pride, their unremitting hard work over the last 30 years has finally yielded fruitful results. By the end of 2019, the wild population of Przewalski's horses in China had been significantly expanded to 714 that fall into six generations, ranking first in the world. Moreover, 19 Przewalski's horses have survived in natural environment that is completely free from human intervention and successfully gave birth to their offspring. The successful reintroduction of Przewalski's horse is a great achievement for China's wildlife conservation and deserves our heartfelt congratulation.

命途多舛的遗鸥自然保护区
Up and Downs of Relic Gull Nature Reserves

母爱·内蒙古鄂尔多斯
Maternal love — Ordos, Inner Mongolia

在内蒙古鄂尔多斯，遗鸥母亲小心呵护着自己两只新生的小生命，充满了深深的爱。

In Ordos, Inner Mongolia, the mother of relic gulls (*Ichthyaetus relictus*) carefully protects her two newborn babies, with deep love.

遗鸥，是直到1971年才被命名的鸥类，有遗落的珍稀鸥之意。我国的遗鸥最早是于1987年4月在内蒙古陕西交界处被发现的。在中国科学院鸟类专家和我本人（当时具体负责国家自然保护区建设工作）的竭力推动下，1998年，在内蒙古鄂尔多斯的陶力庙—阿拉善湾海子这个全球最大的遗鸥繁殖地上，建立起了当时中国（也是世界）的唯一的遗鸥自然保护区。当时，该自然保护区内记录到的湿地鸟类共计83种，而遗鸥数量最多时能达到16000只，繁殖巢数最多时超过3600个。鄂尔多斯遗鸥自然保护区2002年被《关于特别是作为水禽栖息地的国际重要湿地公约》认定为国际重要湿地[1148号]。

后来，由于鄂尔多斯遗鸥自然保护区附近采矿以及来水减少等原因，水面面积开始逐年变小，遗鸥失去了基本的生存条件，陆续迁移到了附近陕西神木县的红碱淖省级自然保护区，使这里逐渐成为全球最大的遗鸥繁殖基地，多时达7000多个鸟巢。每年4月初，全球90%以上的遗鸥都会从越冬地飞到这里。红碱淖之所以能够成为遗鸥新的繁殖地，最主要的原因除保持了稳定的水面之外，湖中有一个四面环水的小岛，它给遗鸥提供了一个安全、安静的环境，使其免受人类和天敌的伤害。

但是，陕西红碱淖省级自然保护区却还挂有省级风景名胜区、国家级水利风景区、4A级旅游景区等多个牌子。在多块牌子并存的情况下，限制、禁止开发的自然保护区显然处于最弱势和最不受重视的地位。近些年，遗鸥又开始有了新的动向，从这里飞到河北康巴淖尔湖以及附近湖淖的遗鸥在不断地增加。红碱淖自然保护区是否又会重蹈覆辙，像鄂尔多斯遗鸥自然保护区一样最终被遗鸥遗弃？这种一块自然保护地多块牌子的情况，其实就是当今中国很多自然保护地的真实写照。

中央提出，要建立以国家公园为主体的自然保护地体系。2018年新一轮国务院机构改革，国家公园、自然保护区、风景名胜区、自然遗产地以及水利、地质、森林、湿地、沙漠公园等自然保护地统由一个资源综合部门——国家林业和草原局来管理，一块自然保护地只能挂一个牌子。多年来困扰中国自然保护"九龙治水"的尴尬局面终于结束了，这对于提升中国的自然保护工作水平大有裨益，也将为像遗鸥自然保护区这类相似的自然保护地摆脱命运多舛的窘境提供坚实的保障。

The relic gull (*Ichthyaetus relictus*), a species of the Laridae family that was not formally named until 1971, implies in its very name that it is a "precious type of bird past down from far-back eras of evolution history". The earliest relic gulls discovered in China were those found at the junction of Inner Mongolia Autonomous Region and Shaanxi Province in April 1987. In 1998, with the joint advocacies of ornithologists from the Chinese Academy of Sciences and me (I was then in charge of the development of national nature reserves), the only relic gull nature reserve in China (also in the world) was established at Taolimiao-Alax Lake in Ordos, Inner Mongolia, which was also the largest relict gull breeding ground in the world. At that time, a total of 83 species of wetland birds were recorded in the nature reserve, with the number of relic gulls reaching 16,000, and the number of breeding nests exceeding 3,600 during peak times. The Ordos Relic Gull Nature Reserve was recognized as wetlands of international importance [No.1148] in 2002 by the *Convention on Wetlands of International Importance Especially as Waterfowl Habitat*.

Later, due to the impacts of mining activities in places surrounding the Ordos Relic Gull Nature Reserve and the decrease of water supply, the size of wetlands within the Reserve began to shrink year by year, hence depriving the relic gull of their basic living conditions and in turn forcing them to move gradually to the Hongjiannao Provincial Nature Reserve in nearby Shenmu County, Shaanxi Province. As a result, the latter by and by became the largest breeding base of relic gulls in the world, where over 7,000 nests can be found during peak times. At the beginning of April every year, more than 90% of the world's relic gulls will fly here from their wintering grounds. The main reason why the Hongjiannao Provincial Nature Reserve can become a new breeding ground for relic gulls is that, in addition to maintaining a stable water surface, there is a small island surrounded by water in the lake, which provides relic gulls a safe and quiet environment that is free of human intervention and dangers of becoming the prey of their predators.

> 鸥群飞来·陕西红碱淖
> A huge flock of gulls — Hongjiannao, Shaanxi
>
> 遗鸥非常珍稀，是地球上极少数东西方向迁徙的鸟类，它们在鄂尔多斯高原上繁殖，几乎是与地球纬度平行地飞往东方的渤海湾越冬。
>
> The rare and precious relic gull is one of the few bird species on the Earth that migrate in the east-west direction. They live and breed on the Ordos Plateau in summer and migrate eastward almost in parallel along the northern latitude of the Earth to spend winter times in the Bohai Bay.

However, the Hongjiannao Provincial Nature Reserve in Shaanxi Province is bearing the titles of provincial scenic spot, national water conservancy scenic spot and 4A tourist attraction at the same time. Under such a circumstance where the place takes on multiple roles, its role as a nature reserve where development is either restricted or totally banned obviously will receive the least attention. In recent years, there has been a new trend for the relic gulls to move away in increasing number to Kangbanner Lake and other lakes in Hebei Province. Will the Hongjiannao Nature Reserve become a new Ordos Relic Gull Nature Reserve and be abandoned by the fowl? Frankly speaking, many nature reserves in China today are faced with the embarrassing situation where they have to assume multiple roles at the same time.

The central government has proposed to put in place the natural protected area system highlighting the central role that national parks play. The State Council launched in 2018 a new round of institutional reform that aims to commission the administration of all types of protected areas — national parks, nature reserves, scenic spots, natural heritage sites, as well as water conservancy, geological, forest, wetland, and desert parks — to a single administrative authority, namely the National Forestry and Grassland Administration. Only one title could be bestowed to any individual protected area. The newly-implemented management system for the natural protected areas puts an end to the embarrassing situation where protected areas were subjected to the administration of multiple agencies. The introduction of this new system will be conducive to an overall improvement in the country's nature protection endeavors, and lay down a solid foundation for the relic gull nature reserves and other protected areas to fulfill their respective conservational roles smoothly.

荒漠孤舟·甘肃敦煌
The lone boat of the desert — Dunhuang, Gansu

在库姆塔格沙漠，生活着比大熊猫更为珍稀的野生动物——野骆驼。在恶劣的极端干旱气候条件下，野骆驼能够长时间不喝水不进食，沙漠中稀少的沙生植物是维持其生命的唯一食粮。据中外科学家们调查，当前全世界的野骆驼不到1000头，且仅存于中国新疆、甘肃及与蒙古国交界的荒漠、戈壁的狭小范围中。

There lives in the Kumtag Desert a kind of wild animal that is even rarer than the giant panda—wild camel (*Camelus ferus*). In harsh, extremely arid climates, the wild camels can go without water or food for long periods of time, with sparsely-grown psammophytes in the desert being the only food that sustains their lives. According to scientific surveys jointly carried out by Chinese and foreign scientists, the total population of wild camels in the world adds up to no more than 1,000, and they are typically found only in the narrow stretch of deserts in the parts of Xinjiang and Gansu that border the Republic of Mongolia.

1 雪地猞猁·新疆阿勒泰
Lynx in the snowfield — Altay, Xinjiang

猞猁为喜寒猫科动物，基本上属于北温带寒冷地区的产物，也可生活于低纬度高山高寒地带，栖息环境多样。

The lynx (*Lynx lynx*) is cold-loving feline that primarily inhabits cold regions in the north temperate zone. Nevertheless, it can also survives in the frigid high mountainous areas in low latitudes, and therefore its habitats are highly diversified.

2 岸边河狸·新疆布尔根河
Beaver on riverbanks — Burgen River, Xinjiang

中国的河狸非常稀少，分布相当狭窄，仅生活在新疆的布尔根河和乌伦古河等地方。

The beaver (*Castor fiber*) is a very rare animal in China, which normally are only found in the narrow stretch of lands along the Burgen River and the Ulungu River in Xinjiang.

3 林缘狼嗥·河北小五台山
A howling wolf at forest margin — Xiaowutai Mountains, Hebei

狼是世上分布很广的凶猛食肉动物，在生态系统平衡中起着关键作用。但由于狼有时会捕食鸡、羊等家禽家畜，免不了遭到人类的捕杀，现在狼已经很稀少了。狼会以凄厉的叫声警示同伴或宣告势力范围，谓之"狼嗥"。

The wolf (*Canis lupus*) is a fierce predator widely distributed in the world, and plays a key role in maintaining balances of ecosystem. But because it sometimes preys on chickens, sheep and other domestic animals, it inevitably becomes the target of hunting by humans. Now the population of wolf is very scarce. The wolf often warns its companions or asserts its ownership to a habitat in piercing mournful cries, so called "howling of the wolves".

雪上赤狐·新疆阿勒泰
Red fox in the snow — Altay, Xinjiang

赤狐分布很广，生性多疑且机敏胆小，在中国古代神话中常以"狐狸精"形象出现。"杀过行为"是指肉食性动物一次杀死远远超过自己食量猎物的行为，狐狸、狼、金钱豹等都有这样的行为。

The red fox (*Vulpes vulpes*) is widely distributed. Being suspicious, alert and timid in nature, they often appear in ancient Chinese mythology in the form of "fox elves". "Overkilling behavior" refers to the behavior of carnivorous animals that killing far more number of preys than they could consume, which is a typical habit of such beasts as foxes, wolves, and leopards.

动感山间·
新疆巴音郭楞
Swift argali on rocks —
Bayingolin, Xinjiang

阿尔泰盘羊是典型的山地动物，它们能在高山裸岩带和悬崖峭壁上奔跑跳跃，来去自如，有垂直迁徙的习性。交配期间，雄性阿尔泰盘羊争偶激烈，巨角相撞响声巨大。

The argali (*Ovis ammon*) is a typical mountainous animal that is capable of running and jumping freely on bare rocks and cliffs. It has the habit of vertical migration. During mating seasons, fierce fighting tends to break out often between male argali, whose giant horns collide against each other, making loud noises.

悬崖上的北山羊·
新疆昆仑山
Ibex on the cliff — Kunlun Mountains, Xinjiang

北山羊具有所有羊中最大的角，攀爬和跳跃能力比谁都强，栖息于高海拔的裸岩和山腰碎石嶙峋地带，和雪豹一样堪称栖居位置最高的哺乳动物。

Among all species of goats, the ibex (*Capra sibirica*) has the largest horn. Being highly apt in climbing and jumping, it lives at high-altitude rugged areas of exposed rock and rocky hillside. Like the snow leopard, it is known as the highest-dwelling mammal.

岩羊的旋律·宁夏贺兰山
Melody of blue sheep — Helan Mountains of Ningxia

在杂乱的戈壁岩石上，岩羊群依然如履平地，一只接一只地陆续跳跃，奏响出一曲荒漠生命的乐章。

On rugged rocks in the gobi desert, a herd of blue sheep (*Pseudois nayaur*) are marching ahead elegantly, as if they were moving on obstacle-free plains. Their parades on dangerous cliffs make up a moving melody of life in the deserts.

❶ 蒙古野驴·新疆卡拉麦里
Mongolian wild asses — Karamaili, Xinjiang

蒙古野驴为典型的荒漠动物，极耐干渴，能在干旱的环境中找到合适的地方用蹄刨坑挖水饮用，冬季主要靠吃积雪解渴。

The Mongolian wild ass (*Equus hemionus*) is a typical desert animal that is extremely thirst-tolerant and capable of spotting the right place in dry deserts to dig for drinking water with their hooves. They normally quench thirst by eating snow in winter times.

❸ 鹅喉羚·新疆吉木萨尔
Goitered gazelles — Jimsar, Xinjiang

鹅喉羚是典型的荒漠和半荒漠地区的动物，在新疆卡拉麦里和普氏野马、蒙古野驴等同域，近年来种群数量下降不少。

The goitered gazelle (*Gazella subgutturosa*), a typical animal that inhabits the desert and semi-desert areas, shares roughly the same geographical sphere with Przewalski's horses and Mongolian wild asses in Karamaili, Xinjiang. Its population has declined notably over recent years.

❷ 塔里木马鹿·新疆夏尔西里
Tarim marals — Xiarxili, Xinjiang

塔里木马鹿是生存于新疆塔里木盆地的马鹿亚种，它们在群山深处丛林草地中生长繁衍，其种群已不足10000只。

The Tarim maral (*Cervus yarkandensis*) is a subspecies of red deer living in Tarim Basin, Xinjiang. They grow and breed in either bushes or grasslands in the depth of mountains, with a current population of less than 10,000.

❹ 高鼻羚羊·甘肃武威
Saiga antelopes — Wuwei, Gansu

国内的高鼻羚羊野生种群现已灭绝，为恢复野外种群，经引种回国，正在进行回归野外的实验和研究，但进展缓慢。

Wild population of the Saiga antelope (*Saiga tatarica*, high-nose antelope in Chinese) has gone extinct in China. In order to restore the wild population, some Saiga antelopes have been imported from abroad. Experiments and researches are under way to test feasibilities for reintroducing them to the wild, with slow progress though.

蒙原羚仔·内蒙古苏尼特左旗
A Mongolian gazelle calf — Sonid Left Banner, Inner Mongolia

蒙原羚（俗称黄羊）是健康草原生态系统的重要组成，现已成为稀有物种，处于弱势生态位，急需保护。

As an important component of healthy grassland ecosystems, the Mongolian gazelle (*Procapra gutturosa*, commonly known as yellow sheep in Chinese) has become a rare and highly vulnerable species at present, calling for urgent measures to put it under protection.

沙漠深处天鹅湖·内蒙古腾格里沙漠
Swan Lake deep in the desert — Tengger Desert, Inner Mongolia

在本区大沙漠的丘间洼地里有不少呈原始状态的湖泊，小天鹅、绿头鸭、赤麻鸭、等水鸟自由自在地在这里生活。在沙丘金色光芒的映衬下，天鹅湖显得格外美丽灿烂。

There are many primitive lakes in the hills and low-lying lands of the great desert, where water birds such as whistling swans, mallards, ruddy shelducks and grebes live freely. Against the golden light of sand dunes, the Swan Lake is particularly beautiful and brilliant.

← 起飞瞬间·内蒙古大兴安岭
Taking off — Greater Khingan, Inner Mongolia

初升的太阳懒洋洋地挂在落叶松的树枝上，冰雪覆盖的冬日分外寒冷，长尾林鸮照样也得出来觅食。

The rising sun hangs lazily on the branches of larch, and the snowy winter is extremely cold. At this time, the ural owl (*Strix uralensis*) still has to come out to look for food.

❶ 展翅飞翔·新疆阿勒泰
Soaring high in the sky — Altay, Xinjiang

大鵟为大型猛禽，以鼠类和鼠兔等为主要食物，也食雉鸡、石鸡、昆虫等，其在维护草原生态系统平衡中具有很大的作用。

The upland buzzard (*Buteo hemilasius*) is a large raptor that feeds mainly on murine and pikas. It also eats pheasants, rock partridges, insects, *etc.*, playing an important role in maintaining the balance of grassland ecosystems.

❷ 王者归来·新疆昆仑山
Return of the lord — Kunlun Mountains, Xinjiang

金雕是北半球最广为人知的大型猛禽，位于自然界食物链顶端，以大中型鸟类和中小型兽类为食。逆光飞翔的金雕浑身金光灿灿，一派王者风范。

The golden eagle (*Aquila chrysaetos*) is the most widely-known large raptor in the northern hemisphere. It sits on the top of the natural food chains and feeds on large and medium-sized birds as well as small and medium-sized beasts. Taken against the sun behind it, this golden eagle in the picture is bathed in a gold halo.

❶ 藏雪鸡·新疆天山
Tibetan snowcock — Tianshan, Xinjiang
藏雪鸡对高山裸岩自然条件有着很强的适应能力，能适应雪中生活，体羽多呈棕褐色。
With strong adaptability to the natural conditions dominated by bare rocks in high mountains, the Tibetan snowcock (*Tetraogallus tibetanus*) adapts well to life in the snow. Its feathers are mostly brown.

❷ 石鸡·新疆天山
Rock partridge — Tianshan Mountains, Xinjiang
石鸡耐干旱，主要栖息于低山丘陵地带的岩石和沙石坡上。
The rock partridge (*Alectoris chukar*) is drought resistant and mainly inhabits on rocks and sandstone slopes in low mountains and hills.

❸ 毛腿沙鸡·内蒙古包头
Pallas's sandgrouse — Baotou, Inner Mongolia
毛腿沙鸡大小似家鸽，但尾长而尖，常成群活动，主要栖息于平原草地、荒漠和半荒漠地区。
The Pallas's sandgrouse (*Syrrhaptes paradoxus*) looks like the pigeon in size, but it has a long and pointed tail and moves around in group. It mainly inhabits in plain grasslands, deserts and semi-desert areas.

❹ 沙蜥·宁夏毛乌素沙地
Toad-headed lizard — Mu Us Sandy Land, Ningxia
沙蜥具有适于荒漠和半荒漠及草原生活的习性，可直接从捕食的蚁类和昆虫中获得所需水分。
The toad-headed lizard (*Phrynocephalus* sp.) has the habit of living in deserts, semi-deserts and grasslands, and can obtain water directly from ants and insects.

争雄·内蒙古白云敖包
Intense contest — Baiyun Aobao, Inner Mongolia
每年的繁殖季节，黑琴鸡都要重选配偶。为了争得雌性的芳心，雄性黑琴鸡在"斗鸡盘"上集结后，都要使出浑身解数捉对斗狠，直斗得羽毛乱飞甚至血肉淋漓。在一旁的雌性黑琴鸡会用挑剔的眼光观看、甄选，认准了它的如意郎君后，便以身相许，双飞双宿。
Every year in the breeding season, the black grouse (*Lyrurus tetrix*) has to choose a new mate. In order to win the female's heart, male black grouses would gathered on the "cockfight terrace" for highly competitive contests, in which each of them has to demonstrate their fullest competence until both end up in serious wounds. While the fight goes on, the female bird will stand by and keep a close watch, waiting for her time to mate with the one who prevails in the contest.

生态思考 | ECOLOGICAL REFLECTION
对草原网围栏的思考
My Thoughts on the Use of Fences on Grasslands

撞在铁丝网上的鸟·内蒙古锡林郭勒
A bird caught up in the barbed wire — Xilin Gol, Inner Mongolia

灵活机敏的鸟在网围栏前也难逃厄运。
Agile and vigilant as the birds are, they fall frequently victim to barbed wire set up by human beings.

20世纪80年代以来，在草原管理上采用了草场分包到户、用网围栏划定界限的办法，这种管理方法取得了不少成效，但带来的副作用也是明显的。大片的草原、荒漠草原包括沙地，凡是有植被的地方，大部分都被无数的、横七竖八的网围栏分割成小块的独立王国，呈现出"多、密、乱"的特点。现在，这种管理方法带来的严重生态问题已经不断地显现出来：草场退化，植被盖度减少，定居点四周荒漠化程度加剧，草原载畜量锐减，草原生物多样性大大降低。

由于野生动物的繁衍、迁徙、逃生的通道被阻断而无法正常生存，它们或因无路可逃而被狼等天敌轻易捕获，或因慌不择路而直接撞伤、撞死在铁丝网前。一些试图跳起来翻越围栏的野生动物，每一次跳跃都有可能命丧黄泉。致密的网围栏使岩羊、鹅喉羚、高鼻羚羊、普氏原羚、藏羚、蒙古野驴、蒙原羚等动物甚至包括各种鸟类的活动受到极大程度地限制，撞死、挂死的现象也时有发生。有调查证明，从2003年到2012年，9年的"围栏效应"已经导致新疆北部鹅喉羚种群数量下降50%以上，内蒙古草原的蒙原羚基本消失。中国相当稀缺的高鼻羚羊也因为网围栏的阻隔使物种没法交流而灭绝。网围栏已经成为草原野生动物致命的、无法逾越的障碍。

为了解决这些实际问题，为了通过尽量恢复轮牧来保护草场，近年来，部分地方有些牧民联合起来，自发地拆除网围栏，如青海玉树、内蒙古锡林郭勒、呼伦贝尔的牧民采取"合作社"的形式合并草场、拆除网围栏等。青海湖畔的普氏原羚因为网围栏遇难事件经媒体报道后，引发社会大讨论，公众纷纷呼吁拆除网围栏，当地随即逐步采取了去铁刺、降低围栏高度、减少网围栏密度、留出一定的野生动物迁徙通道等办法，取得了不错的保护效果。这些做法，也得到了中央新闻媒体的充分肯定。

其实，只有完整地保护草原生物多样性、保护特殊的草原生态系统，为草原野生动物提供繁衍、生息的机会，才会有草原的健康恢复和人们美好的未来。从建设国家北方生态安全屏障的高度出发，重新认识草原生态系统的特殊价值，重新审视当今施行的草场网围栏模式的利弊已经成为当前不可回避的重要任务。

For better management of grasslands, a household contracting system has been adopted since the 1980s, under which grasslands are contracted to individual households, with boundaries between that of each household typically marked by fences. For sure, the adoption of such system has played a constructive role in improving the efficiency of management and productivity, but the negative effects are also obvious. Grasslands, desertified grasslands, sandy lands, and in fact almost all vegetated-lands that formerly formed an integrated ecosystem have now been divided into numerous fragmented pieces that are often mutually exclusive to each other. As a result, countless fences sprang up densely and disorderly across the grasslands. Now, the serious ecological problems caused by this approach of management have risen increasingly, leading in consequence to grassland degradation, reduced vegetation coverage, intensified desertification around human settlements, reduced livestock capacity in the grasslands, and greatly reduced grassland biodiversity.

Blockades set up in the breeding, migration and escape routes of wildlife make it hardly possible for the animals to survive. Deprived of escaping routes, the entrapped animals tend to fall easily into the victims of wolves and other predators, or crash themselves in desperation to the fences and meet their bitter deaths. For the animals, each time they attempt to jump over the fences means a new challenge that is highly likely to cost their lives. The dense fences have greatly restricted the activities of

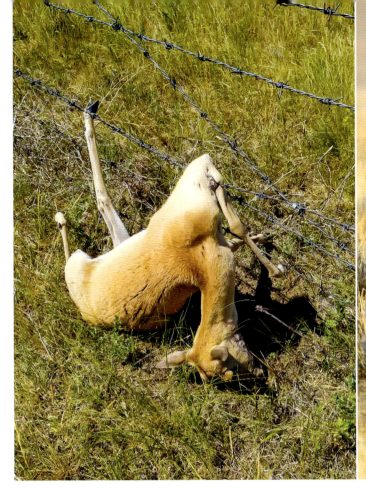

the animals such as the blue sheep (*Pseudois nayaur*), goitered gazelle (*gazelle yarkadensis*), Saiga antelope (*Saiga tatarica*), Przewalski's gazelle (*Procapra przewalskii*), Tibetan antelope (*Pantholops hodgsonii*), Mongolian wild ass (*Equus hemionus*) and Mongolian gazelle (*Procapra gutturosa*). In addition, cases when various birds get caught up in the fences and hung to death on barbed wire are also frequently seen. Studies show that, as a result of the "fence effect", the population of the goitered gazelles in northern Xinjiang has dropped by over 50%, and the Mongolian gazelle (*Procapra gutturosa*) in Inner Mongolia grassland has basically disappeared from 2003 through 2012. The Saiga antelopes, which is a very rare species in China, has also been wiped out by fences that block the communication between species and have become a deadly and impassable barrier for the survival of wildlife on the prairie.

In recent years, in order to solve these problems and protect the grassland by restoring the rotational grazing as much as possible, in some places, some herdsmen united to dismantle the fences spontaneously. For example, herdsmen in Yushu of Qinghai Province, and in Xilin Gol and Hulun Buir of the Inner Mongolia Autonomous Region have taken initiatives in organizing into "cooperatives" that are aimed to merge the grasslands and dismantle the fences. Following a news report about a *Procapra przewalskii* was caught up in a fence and died by the Qinghai Lake, a heated debate among the pubic broke out, calling for the removal of the fences. Then, the local government responded quickly and adopted a series of measures, such as removing iron thorns, reducing the height and the density of fences, and setting aside certain areas as wildlife migration corridors, all of which have been proven to be sound ways for wildlife protection and highly praised by the country's national news agencies.

Frankly speaking, only by completely protecting the biodiversity of the grassland, by protecting the special grassland ecosystem, and by providing opportunities for the reproduction and survival of wild animals inhabiting the grasslands, can the health of the grassland ecosystem be restored and people have a better future. Judged from the height of safeguarding the national eco-security in northern China, it has become an urgent task for us all to re-assess the special value of grassland ecosystems, re-examine advantages and disadvantages of the current fence-based grassland management system that has prevailed over the past few decades.

生存挑战·青海青海湖
Challenges for survival — Qinghai Lake, Qinghai

普氏原羚在奔跑追逐时，常常遇到这样的生死考验，包括求偶时也不例外。

Przewalski's gazelles (*Procapra przewalskii*) are often confronted with life-or-death challenges like this as they play and chase after each other unwarily, and even in the period of courtship.

挂死在铁丝网上的蒙原羚·内蒙古呼伦贝尔
A Mongolian gazelle hanging dead on the barbed wire — Hulun Buir, Inner Mongolia

在铁丝网前，蒙原羚这样的悲剧依然不时发生。

Such tragedies that Mongolian gazelles (*Procapra gutturosa*) run into barbed wire and got hanged still happened from time to time.

雪山、冰川及河源·西藏唐古拉山 世界屋脊终年冰雪覆盖,是中国乃至亚洲的几乎所有大江大河的发源地,被誉为"万山之祖""亚洲水塔"。
Snow mountains, glaciers and the river source — Tanggula Mountains, Tibet The roof of the world is covered with ice and snow all the year round. It is the originating place of almost all the big rivers in China and even Asia. It is called "the ancestor of all mountains" and "Asian Water Tower" in China.

珠穆朗玛峰晨曦·西藏定日
Dawn of Mount Qomolangma — Tingri, Tibet
世界第一高峰，建有以保护极高山、山地森林和灌丛草原生态系统为主的珠穆朗玛峰国家级自然保护区。
Mount Qomolangma is the highest mountain in the world. Mount Qomolangma National Nature Reserve has been set up to protect the ecological systems of extremely high mountains, mountain forests and bush grasslands.

高原雪后·青海大通
High plateau after snow — Datong, Qinghai
冬日晨曦中的村庄点缀在苏醒了的高原面上。
Villages dot on the snow-covered plateau that has just waken up in the morning sunshine on a winter day.

辫状河流·西藏日喀则
Braided rivers — Xigaze, Tibet
雅鲁藏布江中游，江水在河滩上自由自在地流淌着。
In the middle reaches of the Brahmaputra River, water flows freely on riverbeds.

青藏区 包括青海（准噶尔盆地除外）、西藏（东南部除外）和四川西部，是东由横断山脉北端，南由喜马拉雅山脉，北由昆仑山、阿尔金山和祁连山（南坡）各山脉所围绕的区域，属世界动物地理区划中古北界的中国西部分。该区气候带为高原中温带、高原寒温带、高原亚寒带和高原寒带。这里平均海拔4000米以上，有"世界屋脊"之称。动物区系成分主要为高地型，物种以耐高寒的种类为主。由于特殊的地理条件，这里有诸如藏羚、野牦牛、藏野驴、雪豹、鹫等青藏高原特有的大型动物，其中，藏羚是青藏高原具有代表性的旗舰物种。

20世纪80年代，西方世界发现了藏羚羊绒披肩（沙图什）的特殊价值，随后它作为奢侈品在欧洲市场出售，高额的利润促使犯罪分子铤而走险，并发展成了大规模的武装盗猎和非法走私，每年有大约2万只藏羚惨遭杀害。至90年代，我国野生藏羚数量急剧下降，使这一珍贵物种接近灭绝的边缘。1999年，"中国西宁藏羚羊保护及贸易控制国际研讨会"在青海召开并发布了《关于藏羚羊保护及贸易控制西宁宣言》，宣告了国际间合作打击盗猎藏羚、制止藏羚羊绒制品非法贸易活动的机制和行动框架形成，对保护藏羚资源起到了极大的促进作用。更为重要的是，中国政府将藏羚列为紧随大熊猫之后的15个野生动植物拯救工程之一，并在藏羚重要分布区先后划建了西藏羌塘、青海可可西里、三江源和新疆阿尔金山等多处国家级自然保护区，成立了专门的保护管理机构和执法队伍，严厉打击盗猎行为，定期巡山并对藏羚种群活动实施监测。经过一段时间的艰苦努力，以藏羚为代表的高寒野生动物受到了严格的保护，种群和数量也都得到了较快恢复，藏羚由20世纪90年代的6万余只恢复到目前的30万余只，藏野驴也由5万多头恢复到目前的8万多头。

2017年，中国的第一个国家公园——"三江源国家公园"（试点）挂牌成立，大大促进了该区自然保护工作的进一步提高。该区是中国人为活动最少的区域，更是自然保护地占比最大的区域[西藏、青海的自然保护区面积分别占到全自治区（省）国土面积的34.5%和31%]。在此基础上，建议以珠穆朗玛诸峰、雅鲁藏布江大拐弯（含墨脱）、昆仑山、青海湖生态系统为核心建设国家公园。随着当前中国自然保护新时代的开启，以"三江源国家公园"建设为龙头的青藏高原自然保护地体系的建立，将使这块有"亚洲水塔"之称、处于中国乃至亚洲最高生态位的"荒野"，成为全球高原野生生物的最佳乐园。

The Qinghai-Tibet Region (QTR) includes Qinghai (except the Junggar Basin), Tibet (except the southeast) and the western Sichuan, and it is a region surrounded by the northern Hengduan Mountains in the east, the Himalayas in the south, and the Kunlun Mountains, Altun and Qilian (south slop) Mountains in the north. In global zoogeographic zoning, the QTR belongs to the western part of the Palaearctic Realm within China. In terms of climate, this region falls into plateau middle temperate, plateau cold temperate, plateau subfrigid and plateau frigid zones. With an average altitude of more than 4,000 meters, it is known as the "roof of the world". Wild animals inhabiting this region are mainly of the cold-tolerant alpine type. Due to the special geographical conditions, there are such large animals as Tibetan antelopes, wild yaks (*Bos mutus*), Tibetan wild asses, snow leopards (*Panthera uncia*), vultures and so on. Among them, the Tibetan antelope is a representative flagship species in Qinghai Tibat Plateau.

In the 1980s, the western world discovered the special value of Tibetan antelope cashmere shawls (shahtoosh), which were then sold as luxury goods in the European market. Driven by this highly profitable trade, large-scale armed poaching and illegal smuggling of the animal gradually emerged, with up to 20,000 Tibetan antelopes killed each year. By the 1990s, the number of wild Tibetan antelopes in China had declined so sharply that this precious species was almost on the verge of extinction. In 1999, the International Workshop on the Conservation of and Control of Trade in Tibetan Antelopes was held in Xining, Qinghai Province, at which the *Xining Declaration on the Conservation of and Control of Trade in Tibetan Antelope* was formally adopted, declaring the beginning of international cooperation in cracking down illegal poaching of and trading in the animal and its cashmere derivative products. More importantly, the Tibetan antelope protection project was listed among the 15 Key Wildlife Rescue Projects initiated by the Chinese government, whose priority ranks next on to that of the giant panda. National nature reserves were consecutively set up in areas where Tibetan antelopes are densely distributed, such as Qiangtang Nature Reserve in Tibet, the Hoh Xil National Nature Reserve and the Sanjiangyuan National Nature Reserve in Qinghai and Altun Mountains National Nature Reserve in Xinjiang. A special regulatory agency and law enforcement team was also set up to crack down on poaching, patrol the mountains regularly and monitor the activities of Tibetan antelopes. All these efforts led to a rapid recovery of the population of both Tibetan antelopes and other cold-tolerant wild animals. The number of Tibetan antelopes soon recovered from slightly over 60,000 in the 1990s to more than 300,000 at present, and the number of Tibetan wild asses has also increased from just over 50,000 to more than 80,000 at present.

In 2017, the Sanjiangyuan Pilot National Park, the first of its type in China, was formally unveiled and established, marked a major leap forward in the conservation efforts of the QTR region. The least interfered by human activities, the QTR is also the area that has the highest proportion of protected areas in China (with the protected areas in Tibet and Qinghai accounting for up to 34.5% and 31% of their total areas respectively). It is recommended that, building on the above-mentioned achievements, one more national park should be established to cover the ecosystem of Mount Qomolangma, the Great Bend of Yarlungzangbo River (including Motuo) , Kunlun Mountain as well as Qinghai Lake under effective protection. Following the dawn of a brand new era in China's nature conservation efforts, as showcased by the establishment of the Qinghai-Tibet Plateau natural protected area system with the Sanjiangyuan National Park at its core, the QTR — the Asian Water Tower as well as the highest ecological niche in China and Asia — will be a paradise for high plateau-dwelling wildlife species throughout the world.

↑ 攻击前的警示·青海海西
Warning before attack — Haixi, Qinghai

野牦牛为青藏高原特有种，极耐寒，个头庞大且凶猛善战。性情凶狠暴戾的孤牛常会主动攻击在它前面经过的各种对象。野牦牛发起攻击时首先会竖起尾巴示警。

The wild yak (Bos mutus), as the endemic species of Qinghai Tibet Plateau, is a fierce animal with a large size and good at fighting. The lonely fierce wild yak often launches active attacks at other animals passing by. Prior to its attacks, the wild yak tends to raise its tail as a warning.

心系普氏原羚
Heart-felt Concern for Protection of Przewalski's gazelle

发情·青海青海湖
Oestrus — Qinghai Lake, Qinghai

冬天是普氏原羚发情的季节，被雄性普氏原羚追赶得气喘吁吁的母羚正回头张望。

Winter is the season of oestrus for the Przewalski's gazelle (*Procapra przewalskii*). In this photo, a panting female Przewalski's gazelle that is being chased by its male partner turns her head back, looking around vigilantly.

争雄·青海青海湖
Competing for supremacy — Qinghai Lake, Qinghai

普氏原羚一贯性情温和，但为了繁衍子孙、延续基因，雄性之间的争斗也时有发生，而激烈到这种状态的十分罕见。

The Przewalski's gazelle is normally of gentle nature, but fighting occasionally breaks out among males for the right of mating and gene inheritance. It is a rare event for such fighting to go as intense as the one caught in this photo.

普氏原羚又称中华对角羚，中国特有种，曾广泛分布于内蒙古、宁夏、甘肃及青海等地，是较典型的荒漠、半荒漠动物，奔跑时像离弦之箭，不时地在空中跳跃，划出一道道起伏的曲线，分外优美。当凛冽的北风让青海湖上冻时，性情温和的普氏原羚就进入了发情季节，雄羚发出阵阵高昂的咩叫，不时在求偶场呼唤和追逐雌羚。雄性有争偶现象，一般说来争斗并不十分激烈。由于人类活动的影响及栖息地的恶化，该物种数量急剧下降，分布区范围锐减，现只生活在青海湖一带及近祁连山谷的狭小地域，到20世纪90年代已经不到300只，是世界上最濒危的有蹄类动物。

普氏原羚的种群一度受到的威胁极大，主要有：栖息地的缩小、破碎化和沙化，水源缺乏，家畜超载争食，网围栏的阻隔等。经过多年不懈的保护努力，青海湖自然保护区通过强化管理、修建饮水池、预留迁徙通道、降低或拆除网围栏、投喂越冬草料等方式保证普氏原羚的生存，普氏原羚的栖息环境得到了大大的改善。2018年，普氏原羚数量已达到2793只，种群栖息地从7个扩展到13个，创历年新高。普氏原羚的拯救是国家林业局2000年提出的15个野生动植物拯救工程之一，全球都很关注。近20年来，我无数次到青海湖了解考察普氏原羚的保护，目睹了濒危边缘的普氏原羚经挽救逐步恢复种群的一系列过程，甚为欣慰，还居然拍到了极为罕见的普氏原羚激烈争雄并交配的镜头，非常感谢如此眷顾我的普氏原羚！

The Przewalski's gazelle (*Procapra przewalskii*), also known as "Chinese diagonal antelope", is an indigenous species that used to be extensively distributed in Inner Mongolia, Ningxia, Gansu and Qinghai provinces in China. It is a typical desert and semi-desert animal that excels in running and jumping at extremely fast speed with graceful poise. As the Qinghai Lake gets frozen under the influences of biting northern winds each year, the mild-tempered Przewalski's gazelle also ushers in their mating season, with loud courting bleating from male Przewalski's gazelle heard from time to time. Fighting sometimes breaks out among male Przewalski's gazelle for the right of mating, but normally the fight is not intense. Impacts of human activities, coupled by worsening habitat deterioration, have led to a sharp decline in both the population of the animal and the areas where they are distributed. Now they are found only in the narrow stretch of lands near the Qinghai Lake and the valley of the Qilian Mountains. By the 1990s, the population of the animal had dropped below 300, making it one of the most endangered ungulates in the world.

For a period in the past, the survival of Przewalski's gazelle was seriously threatened, by such causes as fragmentation and desertification of ever-shrinking habitats, water shortage, competition from domestic livestock for food, and obstruction of life corridors by the existence of fences. Thanks to years of unremitting conservation efforts made by the Qinghai Lake Nature Reserve to safeguard the survival of Przewalski's gazelle, which include strengthening management, building drinking pools, setting aside migratory corridors, lowering/removing fences, as well as supplying extra forage for the animal to overwinter, the habitat environment of Przewalski's gazelle has been greatly improved. In 2018, the population reached 2,793 and the number of their habitats increased from seven to 13, a record high. The Przewalski's gazelle Rescue Project was among the 15 Key Wildlife Rescue Projects proposed by the State Forestry Administration in 2000 which attracted worldwide attention. In the past 20 years, I have visited Qinghai Lake countless times to investigate on the protection of Przewalski's gazelle, and have personally seen the process through

which the Przewalski's gazelle has evolved from the state of close to extinction to a stage of gradual recovery in population. Even more rewarding is that I have had the chance to capture the rare and precious scene when male Przewalski's gazelle compete fiercely for the right of mating through my camera lens. I owe the Przewalski's gazelle for favoring me like this.

半河清水半河鱼
Crystal-clear Rivers Teeming with Fishes

青海湖是中国的第一大湖，作为维系青藏高原东北部生态安全的重要水体，不仅是控制西部荒漠化向东蔓延的天然屏障，更是青藏区生物多样性较为丰富的重要地区。

青海湖裸鲤（俗称湟鱼）是青海湖的特有鱼类，每年5~8月是青海湖裸鲤洄游的季节，湖内即将产卵的鱼开始在湖边各大河流的入水口集结，它们相约成群溯流而上，穿过怪石嶙峋的河底，聚集在河水相对平缓的地方稍事休息，积蓄力量后再继续前进，直至找到合适的地方排卵受精。由于河道里鱼的密度很大，产生"骑马涉水踩死鱼"的景象并不夸张。在此时段，棕头鸥也不会轻易放过这个年度饕餮大餐，它们大量集聚在河道附近抓捕河中的美食，会形成"半河清水半河鱼""群鸟猎鱼""鱼跳龙门"等现象，堪称世界奇观。

20世纪50年代，为了解决农场耕地的灌溉用水，几条通往青海湖的河流上修筑了大坝，大坝的修建破坏了湟鱼的天然产卵通道，大量上溯的鱼群聚集在坝下直至死去。再加上多年来的过度放牧、开垦和干旱少雨等因素的影响，青海湖流域生态环境一度呈恶化的趋势，青海湖裸鲤资源锐减。为了解决这些问题，20世纪以来，政府对于青海湖流域周边地区的生态环境下大力气进行综合治理，拆除了部分拦河坝并修（改）建了多座台阶式的洄游通道，解决了青海湖裸鲤的洄游问题，其数量开始得到不断恢复。除了自然繁育的青海湖裸鲤外，近十几年来，渔业主管部门每年都会人工培育鱼苗，进行人工增殖放流。截至2017年年底，青海湖裸鲤由2002年的0.26万吨恢复到8.12万吨，是保护初期的31倍。我衷心地希望，"半河清水半河鱼"和"群鸟猎鱼"的奇观不仅是现在而是永远都在青海湖畔呈现。

Qinghai Lake is the largest lake in China. As an important water-body that sustains the ecological system of the northeastern part of the Qinghai-Tibet Plateau, it is not only a natural barrier that contains the eastward spread of deserts in West China, but also an important area that boasts a rich diversity of lives in the Qinghai-Tibet Region (QTR).

The naked carp (*Gymnocypris przewalskii*, commonly known as Huang Yu in Chinese) is a unique fish found only in the Qinghai

守株待兔·青海青海湖
Wait for windfalls —
Qinghai Lake, Qinghai

溯流而上的青海湖裸鲤促成了棕头鸥的一场年度饕餮盛宴。
Brown-headed gulls (*Chroicocephalus brunnicephalus*) are waiting for their annual grand feasts — naked carps that migrate upstream from the Qinghai Lake to their spawning places.

溯河洄游·青海青海湖
Anadromous migration —
Qinghai Lake, Qinghai

在并不深的河流中，成群结队溯流而上的青海湖裸鲤，穿过怪石嶙峋的水体，冲过天敌的层层阻拦，为了找到一个产卵的地方繁衍后代而前赴后继。

Huge flocks of naked carp (*Gymnocypris przewalskii*) gather in a shallow river to get ready for migrating upstream. They would have to carefully navigate the rocky river bottoms, evade lurking predators before reaching suitable places for spawning.

Lake. The period between May to August every year marks the time when the naked carps in Qinghai Lake will migrate to rivers for spawning. Huge flocks of fish gather at the inlets of major rivers that feed the Lake, waiting for the right time to swim upstream for spawning. They carefully navigate the rocky river bottoms and stop for rest at places where the water is relatively calm before swimming further upwards along the rivers, until they reach a suitable place to give birth to their offspring. The population density of fish in the rivers is so high that it is by no means an exaggeration that "fishes are often trodden to death by horses that happen to be crossing the rivers". In this period, huge crowd of brown-headed gulls (*Chroicocephalus brunnicephalus*), unwilling to let their annual chances for a grand feast slip by easily, will also gather in river courses to gobble their delicious preys. Hence come into sight the amazing scenes that are often aptly depicted as "crystal-clear rivers teeming with fishes", "birds feasting on fish", and "flying fishes over the water". Indeed a spectacular wonder of the world!

In the 1950s, to meet the needs of farmland irrigation, dams were built on some major rivers that feed the Qinghai Lake, which obstructed the natural spawning channels of the naked carps and left huge amount of fish that had migrated upstream to the dams unable to go further and died in the end. To make things worse, years of overgrazing, reclamation, drought, scarce rain and other factors have resulted in further deterioration of the ecological environment in the catchments of Qinghai Lake, leading in turn to a sharp drop in the naked carp resources in the Lake. In order to solve these problems, since the 20th century, the government has made great efforts to comprehensively control the ecological environment of areas in the catchments of Qinghai Lake. Some dams that blocked the migratory corridors of fishes have been duly dismantled, which is further complemented with the construction of terraced migratory passages to facilitate their migration. As a result, the population of naked carps in the Qinghai Lake gradually recovered. In addition to help with the recovery of nature-bred naked carps in the Qinghai Lake, the fishery authorities have also been introducing artificially-bred fish to the Lake to enrich the fish resources over the past decade. By the end of 2017, naked carps outputs in the Lake had risen from 2,600 tons in 2002 to 81,200 tons, marking a 31-times increase as against that prior to the implementation of the protective measures. I sincerely hope that the spectacular view characterized by "crystal-clear rivers teeming with fishes" and "birds feasting on fish" will become an ever-lasting wonder of the Qinghai Lake.

家园守护者·西藏日喀则
Guards of homeland — Xigaze, Tibet

希夏邦马峰是唯一一座完全在中国境内的海拔8000米以上的高峰，藏语是"气候严寒、天气恶劣多变"之意，山下的荒漠草原却是藏野驴的天堂。四头藏野驴像站岗卫兵一样有序地横排着，好像在告诉人们：不要进来，这里是属于我们的。

Shishapangma is the only mountain entirely located within China's territory that rises up to an altitude of more than 8,000 meters. It means "extremely cold and changeable weather" in Tibetan language. The desert grassland at the foot of the mountain is a paradise for Tibetan wild asses (*Equus kiang*). Four Tibetan wild asses are arranged in an orderly way like guards, as if warning intruders to stay away from their territory.

雪豹·青海格尔木
← Snow leopard — Golmud, Qinghai
雪豹因常栖息在高原雪线附近而得名"雪山之王",是世界分布海拔最高的大型猫科动物,也是高山生态系统的伞护种。青藏高原及帕米尔高原是雪豹的主要分布区。
The snow leopard (*Panthera uncia*), known as "king of snow mountain", mainly inhabits near the snow line on high plateaus. It is the highest-dwelling big cat in the world and the umbrella species of the alpine ecosystem. Qinghai-Tibet Plateau and Pamir Plateau are the main distribution areas of snow leopards.

① 西藏马鹿·西藏桑日
Shou cervus — Sangri, Tibet
西藏马鹿主要群体仅分布在西藏山南地区的桑日县一带,数量极为稀少,国际上曾一度认为已经灭绝。
The main community of Shou cervus (*Cervus wallichii*) is distributed only in Sangri County, Shannan Prefecture of Tibet. With an extremely small population, this animal was once believed to have gone extinct in the world.

③ 藏原羚·青海可可西里
Tibetan gazelle — Hoh Xil, Qinghai
在西藏有蹄类动物中数量最多、分布最广的是藏原羚,大概是因为它们更能适应被人类改变了的生境吧。图为藏原羚母与子。
The most abundant and widely distributed ungulates in Tibet are Tibetan gazelles (*Procapra picticaudata*), presumably because they are more adaptable to the habitats that have been altered by humans. The mother and her son are captured in the photo.

② 白臀鹿·青海祁连山
McNeill's deer — Qilian Mountains, Qinghai
白臀鹿因臀部有大面积的黄白色斑而得名,主要生活在青藏高原边缘地带。
The McNeill's deer (*Cervus wallichii macneilli*) is named for the large yellow-white spots on its buttocks and lives mainly on the edge of the Qinghai-Tibet plateau.

④ 藏鼠兔·青海玉树
Moupin pika — Yushu, Qinghai
藏鼠兔主要栖息于高海拔的林区、灌丛和草坡,经常和褐背拟地鸦、雪雀等"鸟鼠同穴"。
The Moupin pika (*Ochotona thibetana*) mainly inhabits in high-altitude forests, shrubs and grassy slopes, often found to share the same holes with Hume's groundpeckers and snow finches.

◀ 愤怒·青海海南
Rage — Hainan Prefecture, Qinghai

胡兀鹫性孤独，常单独活动，不与其他猛禽合群，嗜食腐肉。

The bearded vulture (*Gypaetus barbatus*) is a solitary bird that often lives alone and keeps away from other raptors, feeding primarily on carrion.

飞离·四川康定 ▶
Taking off — Kangding, Sichuan

秃鹫为青藏高原最有名的大型猛禽，主要以哺乳动物的尸体为食。图为食牦牛肉后秃鹫的起飞瞬间。

The black vulture (*Aegypius monachus*) is the most famous large birds of prey on the Qinghai-Tibet Plateau, mainly feeding on dead mammals. The photo shows the moment when the black vulture started to take off after eating yak meat.

◀ 争斗·青海海西
Confrontation — Haixi, Qinghai

高山兀鹫与大嘴乌鸦为领地而争斗，最终作为大型猛禽的高山兀鹫竟然落荒而逃。

Surprisingly, the Himalayan vulture (*Gyps himalayensis*), a large raptor, would lose the fight in a confrontation against the large-billed crow (*Corvus macrorhynchos*) for territory.

生态思考 | 国际合作应对盗猎危机的成功典范
Ecological Reflection | A Successful Case of International Cooperation in Addressing Poaching Crisis

新生·青海可可西里
Birth of new life — Hoh Xil, Qinghai

青藏高原又恢复了平静，经历了浩劫的藏羚获得了更大的生命期望，新生的小藏羚让我们看到了野生动物保护工作取得的卓越成就和美好前景。

The Qinghai-Tibet Plateau has regained its serenity after a crackdown on illegal poaching that has taken a serious toll on the Tibetan antelope (*Pantholops hodgsonii*). The newborn Tibetan antelope not only testifies to the great achievements we have made in wildlife protection, but also foretells a bright prospect for our future work in this field.

藏羚（原名藏羚羊）是青藏高原动物区系的典型代表，也是青藏高原生态系统的旗舰物种，只栖息在青藏高原腹地海拔4600~6000米的环境中，季节性迁徙是它们重要的生态习性。藏羚在空气稀薄的高原上奔跑，时速可达180千米，常使狼、熊等食肉兽类望而兴叹。但是，藏羚再快也跑不过走私盗猎集团的枪弹，20世纪90年代，由于境外藏羚羊绒及其制品贸易的兴起，高额的利润导致藏羚被大量猎杀并走私出境，其数量从20世纪60年代的100多万只急剧下降到90年代的6万余只。1994年1月18日，青海省玉树藏族自治州治多县西部工作委员会书记索南达杰为保护藏羚，同18名偷猎者枪战，英勇牺牲。

中国政府将藏羚列为15个野生动植物拯救工程之一，并先后在藏羚的重要分布区划建了西藏羌塘、青海可可西里、青海三江源及新疆阿尔金山等多处国家级自然保护区，成立了专门的保护管理机构和执法队伍，定期进行巡山和种群监测。仅2003年"5·9"特大武装盗猎藏羚案，就抓获2个武装盗猎团伙共9人，缴获藏羚羊皮700余张以及小口径步枪、作案用车等盗猎工具。为敦促国际社会加强合作，在中国政府和《濒危野生动植物种国际贸易公约》（简称CITES）秘书处的共同倡导下，1999年10月，"中国西宁藏羚羊保护及贸易控制国际研讨会"在青海召开，来自中、法、英、美、意、印度和尼泊尔的有关政府机构，CITES秘书处以及世界自然基金会等国际组织的代表集聚一堂，我作为中国政府的代表之一，和各方代表一起进行了讨论，会后发布了《关于藏羚羊保护及贸易控制西宁宣言》。这份宣言的发布，标志着国际间合作打击盗猎藏羚、制止藏羚羊绒制品非法国际贸易活动的局面形成，对保护藏羚起到极大的推动作用。

印象尤为深刻的是，2000年8月19日，三江源自然保护区正式成立，由江泽民总书记题写碑名的"三江源自然保护区"纪念碑奠基在长江源通天河畔，在中央电视一台长达3小时的现场直播中，我以专家和行政部门领导的双重身份接受著名主持人白岩松的采访，在现场强烈呼吁国际社会共同采取行动来保护中国的藏羚，还出示了藏羚羊绒围巾——"沙图什"，并首次提出了"生态安全"的概念。之后不久，我被调到CITES中国履约机构做主要负责人，继续在新的工作岗位上为藏羚的保护作出努力。我们充分利用CITES公约的履约机制提出：没有消费就没有杀害。国际社会要加强协作，必须要齐心协力，坚决打断并消灭在西藏盗猎、在印度和尼泊尔粗加工转运、在欧洲再加工并销售的这条罪恶链条。

在国际社会的大力支持下，经过中国政府的不懈努力，藏羚的保护工作终于取得了丰硕的成果——其数量由20世纪90年代的6万余只恢复到目前的30万余只。2016年9月，世界自然保护联盟（IUCN）将藏羚从"濒危"降为"近危"，连续降低了2个级别。藏羚的保护成为了国际社会精诚协作、共同推进物种保护和恢复的最重要的成功典范。

The Tibetan antelope (*Pantholops hodgsonii*), as a typical representative of the fauna of the Qinghai Tibet Plateau as well as the flagship species of the Qinghai Tibet Plateau ecosystem, only inhabits the hinterland of the plateau at an altitude of 4,600 - 6,000 meters, with seasonal migration being their important ecological habit. Tibetan antelopes are capable of running in the thin air on the plateau at a speed of up to 180 kilometers per hour, making wolves, bears and other carnivores powerless in catching them. However, fast as the Tibetan antelopes are in running, they still fall victim to the bullets fired by illegal poachers. In the 1990s, due to the rise of overseas trade in Tibetan antelope cashmere and its derivative products, high profits led to a large number of Tibetan antelopes being poached and smuggled out of the country, and their population dropped sharply from more than 1,000,000 in the 1960s to slightly over 60,000 in the 1990s. On January 18, 1994, Suo Nan Da Jie, secretary of the Western Work Committee of Zhiduo County, Yushu Tibetan Autonomous Prefecture, Qinghai Province, bravely sacrificed his own life in a gunfight against 18 poachers to protect Tibetan antelopes.

The Tibetan Antelope Rescue Project has been listed by the Chinese government among the 15 Key Wildlife Rescue Projects. In the primary regions that Tibetan antelopes inhabit, several national nature reserves have been established, such as the Qiangtang Nature Reserve in Tibet, the Hoh Xil Nature Reserve and the Sanjiangyuan National Nature Reserve in Qinghai Province, and the Altun Mountains National Nature Reserve in Xinjiang. And special protection management agencies and law enforcement teams have also been set up to regularly patrol the mountains and monitor their population. In the "May 9" large-scale armed poaching of Tibetan antelopes case in 2003 alone, nine people in two armed poaching groups were arrested, with over 700 pieces of Tibetan antelope skin, some small caliber rifles, vehicles and other tools used for poaching seized. In order to urge the international community to strengthen cooperation, under the joint sponsorship of the Chinese government and the Secretariat of the *Convention on International Trade in Endangered Species of Wild Fauna and Flora* (CITES), the International Workshop on the Conservation of and Control of Trade in Tibetan Antelope was held in Xining, Qinghai Province in October 1999, in which representatives of relevant government agencies from China, France, the UK, the United States, Italy, India and Nepal, the CITES Secretariat, the WWF and other international organizations participated. As one of the representatives of the Chinese government, I held discussions with representatives of all parties, and after the meeting, "*Xining Declaration on the Conservation of and Control of Trade in Tibetan Antelope*" was issued. The publication of this Declaration marked the formal initiation of international cooperation in combating poaching of Tibetan antelope and stopping the illegal international trade in Tibetan antelope cashmere products, and played a great role in promoting the protection of Tibetan antelopes.

What impressed me particularly was the official launch of the Sanjiangyuan National Nature Reserve on August 19, 2000, when the commemorative monument bearing President Jiang Zemin's hand-written inscription was ceremonially laid down by the Tongtianhe River — the originating place of the Yangtze River. In a three-hour live broadcast featuring the event on CCTV-1, I wearing simultaneously two hats as both a member of expert panel and the representative from competent government agency, accepted the interview of the famous host Bai Yansong. I urged strongly during the interview the international community to take joint action to protect China's Tibetan antelopes. I also showed the Tibetan antelope cashmere scarf, shahtoosh, and put forward the concept of ecological security for the first time. Shortly afterwards, I was transferred to the CITES implementation agency of China as the director general and continued to work in my new position for the conservation of Tibetan antelope. We took full advantage of the CITES implementation mechanism, and highlighted in particular a key notion for the protection efforts — "no trading, no killing". The international community must work together to break and destroy the evil chain that involves illegal poaching in Tibet, primary processing and transshipment via India and Nepal, and reprocessing and distribution in Europe.

With the strong support of the international community and unremitting efforts of the Chinese government, the protection of Tibetan antelope has finally achieved fruitful results, with the population of the animal quickly recovered from slightly over 60,000 in the 1990s to more than 300,000 at present. In September 2016, the International Union for Conservation of Nature (IUCN) downgraded Tibetan antelope from being "endangered" to being "near threatened", down by two levels on the scale of urgency for protection. The protection of Tibetan antelope has become a telling case about the successful cooperation among the international community in promoting species protection and recovery.

玉珠峰下·青海玉树
At the foot of Yuzhu Peak — Yushu, Qinghai

藏羚是高原的精灵，也是这块土地的主人，我们的责任就是要保护好以藏羚为代表的高原生态系统，保护好这片中国生态位级别最高的圣地。

The Tibetan antelope is not only the elf of the plateau, but also the owner of the land. It is our responsibility to keep the plateau ecosystem represented by Tibetan antelopes and the holy land of this highest-lying ecological niche in China safely protected.

云雾金山·云南独龙江
Cloud-shrouded golden mountains — Dulong River, Yunnan

来自孟加拉湾强大的暖湿气流从南、西南方向汹涌而来,喜马拉雅山东段及横断山间常常雾气腾腾、白云缭绕、金山时现。
Strong warm and moist air from the Bengal Gulf surges in from the south and southwest. So the eastern part of the Himalayas and the Hengduan Mountains are often foggy and cloud-shrouded, with golden appearance.

西南区
South-west Region

山峦叠嶂·高黎贡山西坡
Endless high mountains — the western slope of the Gaoligong Mountains

这里山连谷山，林牵着林；这里孕育千种多样的生命；这里是全球不可多得的生物多样性基因库。
The endless mountains and forests here are home to a great variety of lives, preserving precious biodiversity gene bank of the Earth.

原始针叶林·西藏墨脱
Primitive coniferous forest — Motuo, Tibet

由于地带性和海拔差的共同作用，本区北部与青藏区的边界错综复杂，但有一条线是非常清楚的——以寒温带亚高山针叶林分布带为本区上限。
Due to the variations in both zonality and elevation, the boundary between the northern part of this region and the QTR is complicated, but one line is very clear — the upper limit of this region is the subalpine coniferous forest distribution zone in the cold temperate zone.

高山峡谷·云南怒江峡谷
Alpine valleys — Nujiang Valleys, Yunnan

山高谷深，世界著名的"三江并流"就出现在这里。亚洲水塔的水从世界屋脊经过这里倾泻而下，到华中、到华南、到南亚、到东南亚。
With high mountains and deep valleys, this is a place well known for three big rivers running in parallel from the Asian water tower at the roof of the world to the Central China, South China, South Asia and Southeast Asia.

西南区

西南区 北起青海和甘肃的南缘，南抵云南中南部，包括云贵高原主体、岷山、大雪山、横断山、昌都东部以及喜马拉雅南坡针叶林带以下的山地，由于地带性与海拔差对自然地理环境的共同影响，区域边界非常复杂。本区大部分属于高山峡谷的大横断山区，属世界动物地理区划中东洋界的中国西部分。动物区系成分有高地型、南中国型和东洋型，还有独特的喜马拉雅山—横断山区型。气候区为高原温带、高原亚温带、中亚热带和南亚热带。这里是世界上高山峡谷分布最密集的地区，且大多为南北走向。高山部分有利于北方种类的南伸，沿高山可南伸至云南；峡谷部分有利于热带种类的北延，沿谷地可北延入横断山脉的中段，因此该区生物多样性异常丰富。该区是全球34个生物多样性热点地区之一，被誉为古老物种的庇护所和物种的演化中心，是诸如大熊猫、小熊猫及包括川金丝猴、滇金丝猴、怒江金丝猴、高黎贡白眉长臂猿、西黑冠长臂猿、藏酋猴、菲氏叶猴、熊猴和印支灰叶猴等在内的9种灵长类动物的天堂，是中国灵长类物种最丰富的地区，同时也是中国雉类、鹛鹛类和鸦雀类等鸟类的分布中心及可能的起源地。

该区是中国生物多样性最为丰富的地区，拥有全国鸟类、哺乳动物的50%以及高等植物的30%以上的物种，其野生动物的分布特点是种类多、濒危程度高、特有种占比高、有很多冰川期孑遗物种。但是，该区野生动物物种的种群数量小，分布区狭窄，所建自然保护区的面积都不太大，很多自然保护区呈狭长带状，尤显脆弱，一旦遭受破坏，物种就会迅速陷入濒危以致走向灭亡且难以挽回。随着当今社会经济的高速发展，人为活动的增加，道路交通网密度的不断增大，野生动物栖息地的破碎化、岛屿化日益加剧，其生存空间不断被压缩。与其他区相比，这里野生动物保护工作的客观要求高、保护难度大。因此，对于该区保护等级的要求、管理和资金支持的标准也必须高。

大熊猫国家公园的试点，在探索自然保护地保护与发展，跨省行政协调好中央和地方关系，对自然资源分级确权立责等方面，将摸索出一条具有中国特色的自然保护地建设发展之路。该区应该更多地提高自然保护地的严格保护程度，在"自然保护地分级分类分区"管理上狠下功夫，必须实行最高层级的保护战略。建议在大熊猫国家公园建设的基础上，以整体的高黎贡山生态系统建立国家公园，还应该加大如灵长类、雉类等国家公园的建设，加大保护管理的力度，加大科研和监测的力度，加大资金和人力的投入，才能保护住这座中国乃至世界难得的生物多样性宝库。

The South-west Region (SWR) extends from the southern margin of Qinghai and Gansu in the north to the central and southern parts of Yunnan Province in the south, including the main body of the Yun-Gui Plateau, Minshan Mountains, Daxue Mountains, Hengduan Mountains, eastern Qamdo and the mountains below the coniferous forest belt on the southern slope of the Himalayas. Due to the co-influence of highly varied topography features and altitudes on the natural geographical environment, the regional boundary here is extremely complex. Most of the region belongs to the mountainous area marked by the Hengduan Mountains and other high-rising valley, and falls into the Oriental Biogeographic Realm in global zoogeographical zoning. Its fauna composition includes types of highlands, of southern China, of the oriental, as well as those belonging to the unique Himalaya-Hengduan species. The region covers four climate zones, namely: the high-altitude temperate zone, the high-altitude sub-temperate zone, the mid-subtropical zone and the southern subtropical zone. This is the part of the world that has the densest distribution of alpine valleys, most of which goes in north-south direction. Whereas the alpine mountains are favorable for the southward extension of northern species, in some cases extending so far as to certain places in Yunnan Province, the deep valleys favor the northward extension of tropical species, extending so far as to the middle section of the Hengduan Mountains. Therefore, this region is endowed with an extremely high biodiversity. Known as a haven for ancient species and a centre for species evolution, the SWR is one of the world's 34 biodiversity hotspots. It's a paradise to giant panda, red panda (*Ailurus fulgens*) and nine species of primates including Sichuan golden monkey (*Rhinopithecus roxellana*), black golden monkey (*Rhinopithecus bieti*), Nujiang golden monkey (*Rhinopithecus strykeri*), Gaoligong white-browed gibbon (*Hoolock tianxing*), western black crested gibbon (*Nomascus concolor*), Tibetan macaque (*Macaca thibetana*), Phayre's leaf-monkey (*Trachypithecus phayrei*), Assamese macaque (*Macaca assamensis*) and Indochinese silvered langur (*Trachypithecus crepusculus*). It is not only the region with the most abundant primate species in China, but also the distribution center and possible originating place of some indigenous Chinese birds like pheasants (*Phasianus*), laughing thrushes (*Garrulax*) and corvids.

The SWR has the richest biodiversity, boasting 50% of the bird and mammal species, over 30% of the higher plant species of China. The distribution of wild animals is characterized by various species, high degree of endangered status, and high proportion of both endemic species and the species that have been passed down from the ice age. But given that the populations of wild animals here are typically very small and their habitats are often narrow, nature reserves set up for their protection are often not large in scale, forming narrow strips of niches that are extraordinarily fragile and, once destroyed, will lead to irreversible decline, or even extinction of the target species. With the high-speed development of society and economy, increasing human activities and high density of road network, the fragmentation and isolation of wildlife habitats are getting worse. Compared with other regions, the SWR is faced with more challenging situations in wildlife conservation. For this reason, greater technical, managerial and financial supports are requested for conservation work in this region.

Piloting of the Giant Panda National Park will explore for ways towards the development for natural protected areas that have Chinese characteristics in terms of well-coordinated work between the national and local governments during the cross-province administration, as well as in terms of clear confirmation of the rights and allocation of responsibilities of stakeholders with regard to the natural resources at all levels. Great efforts are needed to put in place a tiered and discretionary mechanism for the management of natural protected areas and adopt the most rigorous protective strategies. It is proposed that, through drawing on the experiences learnt from the establishment of the Giant Panda National Park, a national park shall be set up to protect the entire Gaoligong Mountains ecosystem. Besides, efforts must also be further strengthened for the establishment of national parks that feature primates, pheasants and other wild animals, so as to build up our capacity in conservation, management, scientific research and monitoring, as well as in financial and personnel supports. Only in this way can this precious treasure house of biodiversity for China and the world be safeguarded.

↑ 嬉戏·四川卧龙
Having fun — Wolong, Sichuan

大熊猫是中国的国宝，更是中国自然保护的旗舰物种，主要分布在四川西部、陕西秦岭和甘肃南部，在竹林里穿梭采食是它的常态，这样在树上的调皮玩耍却是我们不大认知的另一面。

The giant panda (*Ailuropoda melanoleuca*), China's national treasure and the flagship speices of China's nature conservation, is mainly distributed in the western Sichuan Province, Qinling Mountains in Shaanxi Province and the southern Gansu Province. It's common to see that it shuttles through the bamboo forests foraging for food, but it is indeed uncommon for us to see it playing cutely on trees, as is shown in this photo.

中国国宝　世界宠儿
China's National Treasure, the World's Beloved Ones

尽管美国的粉丝很心塞，但在2019年11月20日，在美国华盛顿国家动物园出生的大熊猫"贝贝"还是按期踏上了返回"祖籍"的旅途。"贝贝"是在美国出生且生活了4年之后回归祖国的，中国和国外的大熊猫科研合作协议有明确的规定，大熊猫的产权始终属于中国，包括大熊猫在国外期间生育的小仔的产权。当然，大熊猫"贝贝"并不是第一个这样做的大熊猫，同样出生在美国华盛顿国家动物园的它的哥哥"泰山"、姐姐"宝宝"已分别于2010年和2017年返回中国。值得我骄傲的是，它们三兄妹的父母"添添"和"美香"被送去美国的时候，我正好担任了中国代表团团长。

那是2001年1月8日，我们中国代表团一下飞机就直奔华盛顿国家动物园，看望先期到达并正在隔离检疫的大熊猫"添添"和"美香"。到了动物园以后我们才知道，已经有一位大人物利用特权看望了还没有展出的大熊猫，他就是当时即将卸任的美国总统克林顿。

1月10日，大熊猫馆正式开馆，有包括美国NBC、ABC、CBS以及英国BBC、日本NHK等全球著名电视台网在内的70多家新闻媒体争相现场报道。我在开幕式的讲话中说道，"添添"的意思是"添丁进口"，"美香"的意思是"美丽芳香"，中国人民祝愿这对大熊猫在美国生活幸福，繁育成功并多生贵仔！后有媒体报道说，大熊猫开馆场面的热烈程度不亚于奥斯卡颁奖典礼，估计来看大熊猫的人数将超过美总统就职仪式的人数（当时小布什即将上任）。更有意思的是，在美媒所做"当前华盛顿最受欢迎的夫妇"的民意调查中，"添添"和"美香"的得票率达到66.7%，远远超过即将上任的小布什总统及其夫人的33%！我特别能够感受到美国人民喜欢大熊猫的感情流露是真真实实的，他们的幽默感也可见一斑。

在长达19年的旅美生涯中，"添添"和"美香"先后孕育了3个儿子和1个女儿（包括2020年出生的第四胎"小奇迹"）。它们每生一仔都会成为当时美国社会关注的热点，不仅各大媒体争相报道，而且参观票很快就被抢购一空，前来观看的民众更是人头攒动、络绎不绝。"宝宝"和"贝贝"的名字也很传奇，是由中美两国的第一夫人彭丽媛女士和米歇尔·奥巴马女士在它们先后满月时一起取的呢！

还有一次我参与的大熊猫活动也非常让人难忘。1997年7月1日，百年耻辱洗去，东方明珠——香港终于回归中国，在这个有重大意义的历史时刻，一对大熊猫作为中央政府的礼物送给了香港，在香港上上下下引起了热烈的反响。香港政府非常重视，指定在香港海洋公园里花巨资单独建设一座大熊猫场馆，并要求必须尽力模拟大熊猫家乡——卧龙自然保护区的自然生态环境，以便让这一对来自四川的"贵客"能够安居香港、生活舒适愉快。在前期工作中，除安排内地的大熊猫专家、饲养管理人员专程到香港考察评估场馆外，香港方面还派了动物、生态、饲养管理等方面最好的专家，以及场馆设计的建筑师多次到卧龙考察、学习、培训，而我参与了其中的筹备工作。大熊猫"安安"和"佳佳"到达香港后，香港市民万人空巷争睹芳容，电视报刊连篇报道，经久不衰。

盘古开天，万物生长，世上珍禽异兽无数，人类历史上还从来没有任何一种动物能像大熊猫那样长时期获得世人如此多的热爱和关切。大熊猫的一举一动让世人魂牵梦绕，它的生长繁育、生活状况、出访旅行总是伴随着无数的故事和企盼。这些年来，大熊猫先后出使到世界19个国家的23个动物园（截至2019年年底）开展大熊猫合作研究并供观赏，深受世界各国人民的喜爱，成为了中外人民之间友谊的使者和桥梁。

其实，这些影响巨大的国际合作活动背后并不简单：是中国政府和人民对于大熊猫保护的充分重视；是自然保护部门几十年来为大熊猫的保护所作出的艰苦努力；是多少基层的科研人员和无数自然保护工作者的无私奉献。野生大熊猫性成熟晚，交配产仔难，繁殖率低且幼仔死亡率高，再加上大熊猫栖息地破碎化，使得大熊猫种群恢复非常困难。20世纪90年代以来，我国实施了"中国保护大熊猫及其栖息地工程"，建立了数十个以大熊猫为主要保护对象的自然保护区和国家大熊猫研究中心，有效地保护了53.8%的大熊猫栖息地和66.8%的野生大熊猫种群。

为了野生动物的保护，20世纪末以来，中国启动了多次的全国陆生野生动物调查，这在世界各国是没有先例的（全国首次陆生野生动物调查中我担任技术总负责人）。在此基础上，还针对大熊猫这一个特殊物种，先后做了4次全国性专项调查。截至2019年年底，野生大熊猫的数量从20世纪70~80年代的1114只增长到1864只，保护的大熊猫栖息地面积从139万公顷增长到258万公顷。中国从2003年开始做人工饲养大熊猫的野外放归，历经16年，已经成功放归11只，其中，成活9只。

中国大熊猫保护的成功是全球濒危物种保护的典范，作为全球野生动物保护的重大成就而载入史册。

2018年10月29日，对于大熊猫保护具有重大里程碑意义的中国大熊猫国家公园管理局正式挂牌成立，该国家公园试点区域面积达2.71万平方千米，涵盖了四川、陕西、甘肃三省几乎所有的60多个大熊猫自然保护区，88%的野生大熊猫种群和70%以上的大熊猫栖息地。值得回味的是，十几年前（2005年），我带的博士生的毕业论文中就重点阐述了在中国大熊猫分布区打破行政界限建立"自然保护区群"的必要性和可行性，其实就是打破自然保护区各自为政、行政割裂、大熊猫生境割裂的状况，要按照重要物种及其生态系统的完整性来保护和管理的自然保护地先期理论探索，就是当前国家公园建设思路的雏形。

Despite the disappointment of American fans, Bei Bei, a giant panda born at the National Zoological Park in Washington, the United States, made his way back to his ancestral home on November 20, 2019. Bei Bei was born in the United States and returned to China after living there for four years. The cooperation agreements between China and foreign countries on giant panda research clearly stipulate that the property rights of the giant panda will always belong to China, including the property rights of the cubs born during the period abroad. Of course, Bei Bei is not the first cub to be returned. His brother Tai Shan and sister Bao Bao were both born at the National Zoological Park in Washington D. C. and returned to China in 2010 and 2017 respectively. What make me proud is that I was the head of the Chinese delegation when their parents, Tian Tian and Mei Xiang, were sent to the US.

It was on January 8, 2001. The Chinese delegation headed directly for the National Zoological Park to visit the giant pandas Tian Tian and Mei Xiang after we got off the plane. They had arrived earlier and were in quarantine. After arriving at the park, we learnt that a big "figure" had already taken advantage of his position to see the pandas before they were ready for the public to look at. This big "figure" was Bill Clinton, the then President of the United States who was soon to come to the end of his term.

The giant pandas were officially put on show for the public on January 10. It attracted more than 70 news agencies including NBC, ABC, CBS, BBC, NHK and other famous network televisions around the world. In my speech at the opening ceremony, I explained the meaning behind the names of the pandas, saying that "Tian Tian" meant "to have more cubs", and "Mei Xiang" meant "beauty and fragrance". The Chinese people wished they would enjoy a happy life in America and breed more cubs! It was reported that the scene on the opening day was almost comparable to the Oscar prize awarding ceremonies. Some people kiddingly said that the number of people that showed up for the occasion could outnumber those that watched the president's inauguration ceremony (George Walker Bush was about to be sworn in at about that time). Interestingly, an opinion poll launched by American media, the Most Popular Couples in Washington D. C., showed that Tian Tian and Mei Xiang received an amazingly high approval rate of 66.7%, far ahead of the 33%-rate for the incoming President George Walker Bush and his wife. I could deeply feel the affection that the American people have for giant pandas and their sense of humor.

During the 19-year life in the United States, Tian Tian and Mei Xiang have given birth to three sons and one daughter (including the fourth cub named "Xiao Qiji" who was born in 2020). It was the hit story of American society when they gave birth. News media swarmed in to report on the events, and tickets sold out immediately. Visitors flooded in to take a look. It is worth mentioning that the names of Bao Bao and Bei Bei were almost legendary. They were given by the first ladies of China and the United States, Peng Liyuan and Michelle Obama, when a special ceremony was respectively held to mark the time when they were one month old.

There is also another unforgettable event featuring the giant panda that I took part in. On July 1, 1997, China wiped out century-old humiliation, and resumed its sovereignty over Hong Kong — the Oriental Pearl. To celebrate this historic moment, a pair of giant pandas were given to Hong Kong as gifts from the central government, which aroused a warm response from the

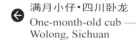

满月小仔·四川卧龙
One-month-old cub — Wolong, Sichuan

在大熊猫的人工繁育工作中，科研人员攻克了"发情难、交配难、育仔难"三大世界难题，为大熊猫野外放归扩大种群奠定了坚实的基础。

In the artificial breeding of pandas, experts have successfully tackled three tough problems facing the whole world, namely late-in-coming sexual maturity, difficult mating and low reproductive rate, which has laid a firm foundation for the reintroduction of pandas to the wild.

顽皮可爱·四川卧龙
Cute and adorable pandas —
Wolong, Sichuan

大熊猫在分类上是熊科动物，而非"猫"，有短短的四肢、圆圆的脸颊、大大的眼圈。其胖嘟嘟的身体憨态可掬，黑白相间的体色异常独特。两只大熊猫只要在一起就特别地顽皮可笑，你推我搡，翻来滚去，尤为惹人喜爱。

Falling among the family of bears rather than that of the felines, giant pandas have short limbs, chubby cheeks and big round eye sockets. The defining features of this animal include their plump bodies and unique black-and-white-intertwining fur colors. The twin pandas that are playing together in the picture, rolling on the ground while pushing and shoving against each other, are particularly adorable and pleasant to the eyes of the viewers.

people of Hong Kong. The Hong Kong government attached great importance to the event and specially allocated a huge budget to build a house for the giant pandas within the Hong Kong Ocean Park. Great efforts were made to make sure that the new built house has the similar natural ecological environment to that of the Wolong Nature Reserve — the pandas' hometown, so that the distinguished guests from Sichuan could lead a comfortable life in Hongkong. During the early stage, the Hong Kong government not only invited giant panda experts and keepers from the mainland to inspect the construction of the house, but also sent carefully-selected experts in zoology, ecology, management and designing to Wolong for thorough training. I had the honor of being a member of the preparatory committee. Upon the arrival of the pandas — An An and Jia Jia, local people of Hong Kong flooded in to see them, which continuously made the headlines of TV and newspapers.

Untold amount of rare and precious wildlife species inhabit the planet of the Earth since it was created, yet no other animal has had the good luck to catch as much attention and love for such a long time from the people around the globe as the giant pandas. People throughout the world follow eagerly on every single event related with the giant pandas — their birth, their health and their wellbeing, their visits to other parts of the world, all these consistently make up interesting stories that touch deep into people's hearts. Over the years, the giant pandas have been sent to 23 zoos in 19 countries (as of the end of 2019) for cooperative research and visit purposes. The giant pandas are so popular around the world that they become the messenger and bridge of friendship between people in China and abroad.

Frankly speaking, all these international cooperative events that have attracted extensive media coverage and produced worldwide influences have not come by easily. Behind these events are the importance that the Chinese government and its people attach to the giant pandas, the decades of painstaking efforts made by natural protection departments, as well as the dedicated work of countless Chinese grassroots scientific researchers and nature conservationists. Recovery of giant panda population is extremely difficult, due to their late-in-coming sexual maturity, difficult mating, low reproductive rate and high mortality of cubs, which is often further worsened by the fragmentation of their natural habitats. Since the 1990s, the Chinese government has initiated the Program for the Protection of Giant Pandas and Their Habitats, through which dozens of nature reserves and national research centers for giant pandas have been established, placing 66.8% of the wild pandas population and 53.8% of their habitats under effective protection.

For the protection of wild animals, the Chinese government has launched a number of nationwide surveys on its terrestrial wildlife since the end of the 20th century, which are unprecedented around the world (I was the chief technical officer for the first nationwide terrestrial wildlife survey). On this basis, China has also conducted 4 national targeted surveys on giant pandas. By the end of 2019, the number of wild giant pandas has increased from 1,114 in the 1970s and 1980s to 1,864, and the area of protected panda habitats has risen from 1.39 million ha to 2.58 million ha. China started in 2003 to re-introduce artificially-bred pandas to the wild. Over the past 16 years, 11 pandas have been put back into the wild among which nine have survived. The success of China's giant panda conservation is a global model of endangered species protection and goes down in history as a major achievement in global wildlife conservation.

The establishment of China Giant Panda National Park Administration on October 29, 2018 marked a historical milestone in China's commitment to the protection of giant pandas. The initial size of the Giant Panda National Park amounts to 27,100 km^2, covering over 60 giant panda nature reserves, 88% of the wild giant panda population and 70% of their natural habitats in Sichuan, Shaanxi and Gansu provinces. In 2005, a doctorate student that I supervised put forward in his doctoral dissertation the necessity and feasibility of setting up "nature reserve groups" that transcend the administrative boundaries in regions where the giant pandas are primarily distributed. In essence, the proposal he raised in the dissertation was meant to put an end to the fragmentation of giant

pandas' natural habitats caused by the lack of coordinated efforts among different nature reserves whose administrative policies and measures are often mutually incompatible. He argued that preliminary theoretical exploration should be made to put into place a natural protected areas administrative mechanism that prioritizes the integrity of ecosystems for critical species, thus envisioning a primitive form of China's national parks that paved the way for our current ongoing efforts in building up national parks.

雉鸡王国
The Kingdom of Pheasants

夫妻随行·云南保山
Faithful couple — Baoshan, Yunan

长尾雉属鸟在我国共有黑颈长尾雉、白冠长尾雉、白颈长尾雉和黑长尾雉4种。图为黑颈长尾雉。
China is home to four species of genus *Syrmaticus*, including black-necked long-tailed pheasant (*Syrmaticus humiae*), Reeves's pheasant (*Syrmaticus reevesii*), Elliot's pheasant (*Syrmaticus ellioti*) and Mikado pheasant (*Syrmaticus mikado*). The one in the picture is the black-necked long-tailed pheasant (*Syrmaticus humiae*).

雉鸡类是古老的鸟类类群，在中国是鸡形目鸟类的通称。在长期的鸟类演化过程中，该类群形成了一些特有的生物学特性，如个体大、地栖、飞翔能力弱、不迁徙、繁殖力较低等，雉鸡类的这些特性使其容易受到环境的影响，因此，它们的生存状况是反映当地生态环境质量的有效指标。栖息地丧失或质量下降，以及人们活动的干扰是雉鸡类受到的主要威胁。

中国是世界上雉鸡类最丰富的国家，有各种野生雉鸡25属64种，拥有世界上近1/4的种类，其中，大约1/3是中国特有种。山鹧鸪属全世界有18种，中国就有10种；马鸡属全世界共有4种，都分布在中国。因此，中国有"雉鸡王国"之称。从近海平原到海拔4000米以上的青藏高原，从东北的林海雪原到海南岛的热带雨林，无论森林、草原、荒漠还是农田生态系统中处处都有雉鸡类生存的足迹。中国现有的自然保护区里，有一半以上数量的自然保护区中都有雉鸡类的分布，很多自然保护区更是以雉鸡类物种为主要保护对象。雉鸡类有超过一半的种类是国家重点保护野生动物，其余的绝大部分被列为《国家保护的有重要生态、科学、社会价值的陆生野生动物名录》。在中国，西南区的雉鸡种类最为丰富，中国雉鸡的大部分种类都可以在此寻觅到踪迹。

Pheasants, an ancient group of birds, is a general term in Chinese used to refer to all the Galliformes. During the long process of bird evolution, the group has developed some special biological characteristics — being large in sizes, living on ground, poor capacity for flying, non-migratory, and of low fertility — that make them vulnerable to environmental influences. Therefore, their living conditions can be taken as very good indicators for the quality of local ecological environment. Habitat loss and degradation, coupled with the interference from human activities, are the primary factors that threaten pheasants.

Pheasants in China have the richest diversity. The country is home to 64 species of 25 genera of wild pheasants and boasts nearly a quarter of the world's total, among which about one third are endemic to China. There are 18 species of partridges (*Arborophila* spp.) around the world and 10 of them can be found in China. And the world has four species of *Crossoptilon*, all of which exist only in China. For these reasons, China is often reputed as the Kingdom of Pheasants. Pheasants are often found in forests, grasslands, deserts and farmland ecosystems that range from the coastal plains to the 4,000m-plus Qinghai-Tibet Plateau, from the snowy forest in northeastern China to the tropical rain forests on Hainan Island. Among China's existing nature reserves, more than half are home to pheasants and many of them feature pheasants as the primary target of protection. Over half of the pheasant species are under priority protection by the state and vast majority of the rest are included in *China's List of Terrestrial Wild Animals under Protection that Are of Significant Ecological, Scientific or Social Values*. The SWR boasts the richest diversity of pheasants, and almost all species of such birds can be found here.

林中精灵·云南高黎贡山
Fairies of the forest — Gaoligong Mountains, Yunnan

白鹇分布较广，非常机警，在茂密的原始林中，白鹇如精灵般稍纵即逝，很难抓拍到。这是在高黎贡山原始湿性常绿阔叶林中首次拍到的白鹇全身照片。中国有鹇属鸟类3种：白鹇、黑鹇和蓝腹鹇。

The silver pheasant (*Lophura nycthemera*) is extensively distributed and alert. In the dense primitive forests, this pheasant would dash by as quick as lightening and it is extremely difficult to capture its movements with camera. This is the first full-body photograph of the pheasant that was taken in the humid evergreen broad-leaved primitive forests of Gaoligong Mountains. China is home to three species of genus *Lophura*: the silver pheasant (*Lophura nycthemera*), the Kalij pheasant (*Lophura leucomelanos*) and the Taiwan blue pheasant (*Lophura swinhoii*).

锦丽之鸡·四川甘孜
Colorful pheasant — Ganzi, Sichuan

白腹锦鸡羽毛色彩艳丽，和红腹锦鸡的羽色相得益彰，在中国传统文化中和红腹锦鸡相同，都是富贵吉祥的象征。

The Lady Amherst's pheasant (*Chrysolophus amherstiae*) has brightly colored feathers, making it a perfect match for golden pheasant (*Chrysolophus pictus*) in color. Both the two species are regarded in Chinese culture as symbols of wealth and fortune.

辉虹之雉・云南怒江
Gorgeous pheasant — Nujiang, Yunnan

图中的白尾梢虹雉与绿尾虹雉、棕尾虹雉共计3种虹雉,以羽毛闪着金属光泽为特征,是喜马拉雅—横断山区特有鸟类,都是数量极少、高海拔分布、极为罕见的大型雉鸡类。

The genus *Lophophorus* characterized by dazzling feathers has three species: Sclater's monal (*Lophophorus sclateri*), Chinese monal (*Lophophorus lhuysii*) and Himalayan monal (*Lophophorus impejanus*). This kind of large-sized and extremely rare pheasants, the genus *Lophophorus*, is the endemic species to the Himalaya-Hengduan Mountains and distributed in areas at very high altitude.

❶ 啼晓之雉・云南保山
Crowing pheasant — Baoshan, Yunnan

"山鹧鸪,啼到晓,唯能愁北人,南人惯闻如不闻。" 唐朝大诗人白居易的诗句,也隐含了山鹧鸪的分布区。图中的白颊山鹧鸪分布区域狭窄,在我国仅分布于云南滇西,数量稀少。我国有世界山鹧鸪属18种中的10种,还有如四川山鹧鸪、白眉山鹧鸪、海南山鹧鸪、台湾山鹧鸪等。

A line in the poem of the famous Tang-dynasty poet Bai Juyi, which goes roughly to the effect that *"whereas the sorrowful crowing of partridges tends to evoke strong emotions for people in the north, the southerners are so much used to it that it looks as if they had heard nothing special"*, implicates the distribution of this fowl in China. The white-cheeked partridge (*Arborophila atrogularis*) shown in this photo has an extremely small population that live only in a narrow stretch of habitats in western Yunnan. Among the 18 *Arborophila* species known to exist in the world, 10 are found in China, typical examples of which include: Sichuan partridge (*Arborophila rufipectus*), white-browed hill partridge (*Arborophila gingica*), white-eared hill partridge (*Arborophila ardens*) as well as the Taiwan partridge (*Arborophila crudigularis*).

❷ 华美之雉・四川唐家河
Splendid pheasant — Tang Jiahe, Sichuan

角雉都有鲜艳华美的外表,我国分布有角雉属的全部5种,分别为红腹角雉、黑头角雉、红胸角雉、灰腹角雉和黄腹角雉。图中红腹角雉主要分布于西南地区。

The tragopan has splendid appearance. China is home to all the five known species of tragopan in the world, namely: the Temminck's tragopan (*Tragopan temminckii*), the black-headed tragopan (*Tragopan melanocephalus*), the crimson horned-pheasant (*Tragopan satyra*), the grey-bellied tragopan (*Tragopan blythii*) and the yellow-bellied tragopan (*Tragopan caboti*). The Temminck's tragopan in this photo is mainly distributed in the SWR.

❸ 血色之雉・四川唐家河
Blood pheasants — Tang Jiahe, Sichuan

血雉是血雉属中的唯一种,栖息于雪线附近,名称来自雄鸟大覆羽、尾覆羽、脚、头侧和蜡膜的血红色。

The blood pheasant (*Ithaginis cruentus*) is the only species of genus *Ithaginis* that mainly inhabits the places situated near the snowline. It's named as such because of the crimson color of the large mulch, tail mulch, feet, head, and wax membrane that is typically found on male birds of this species.

灵长类天堂
The Paradise of Primates

中国是灵长类动物的世界起源中心和物种分化中心之一，而西南区的大横断山就是现代中国灵长类分布的核心地带。中国共有4种金丝猴，本区就分布有除黔金丝猴（华中区）以外的川金丝猴、滇金丝猴、怒江金丝猴3种金丝猴。金丝猴体型在灵长类中属中等个，尾与体等长，毛色以金黄色或黑灰色为主，主要特征是鼻孔与面部几乎平行，因此也称"仰鼻猴"。研究认为，鼻梁骨的退化有利于减少在稀薄空气中呼吸的阻力，是对高原缺氧环境的适应。

长臂猿体型比金丝猴稍小，无尾，因腿短臂长而得名，其手掌比脚掌长，手指关节比脚趾关节长，这是它们在大树之间移动、飞跃都主要依靠长长手臂的原因和结果。雄猿一般为黑色、棕褐色，毛色较深，雌猿或幼猿毛色浅，为棕黄色或金黄色。中国有长臂猿属物种8种（全部为国家一级重点保护野生动物），除海南长臂猿、东黑冠长臂猿、北白颊长臂猿3种长臂猿（华中区）以外，西白眉长臂猿、东白眉长臂猿、西黑冠长臂猿、白掌长臂猿、高黎贡白眉长臂猿5种长臂猿均分布在本区。猿类因其外貌和人类最为相像，科学上称它们为"类人猿"，在血统关系上，它们也与人很相近。猿和猴外形上最显著的区别是，猿没有尾巴、颊囊，而猴有。

叶猴主要以各种树叶、嫩芽为食物，依靠从树叶和幼芽中汲取身体所需的营养和大部分水分来维持生命，因而得名。我国共有7种叶猴，本区有除黑叶猴（华中区）、喜山长尾叶猴（青藏区）和白头叶猴（华南区）3种叶猴以外的印支灰叶猴、菲氏叶猴、戴帽叶猴和肖氏乌叶猴4种叶猴。

人们最常见的猴子就是猕猴了。猕猴的主要特征是尾短体粗壮，有能够储藏食物的颊囊，它们的前肢与后肢大约同长。中国共有8种猕猴，本区生活有除台湾猕猴（华南区）和藏南猕猴（青藏区）之外的普通猕猴、藏酋猴、短尾猴、熊猴、北豚尾猴和白颊猕猴6种猕猴。

本区最值得一提的是，2010年发现的金丝猴新种——怒江金丝猴，2015年发现并由中国科学家定名的新猴种——白颊猕猴，2017年由中国科学家新命名的长臂猿——高黎贡白眉长臂猿都只生活在这里。有科学家推断，今后这个地区还有可能发现灵长类新物种。这里真是一座名副其实的灵长类天堂！

这些灵长类动物都栖息在森林生态系统中，以树栖生活为主。它们多是社会性动物，其生活和迁徙都是成群结队进行

的，除了仰鼻猴属、长臂猿属的动物是以家庭为单位之外，其他猴群大都是有猴王带领的。

中国的所有灵长类动物都非常珍稀，全部都被列为国家一级或二级重点保护野生动物。在我几十年的野生动物保护生涯中，大都在野外见过，其中最难寻觅的恐怕是怒江金丝猴了。在高山峡谷、坡陡林密的怒江峡谷中，我们顶风冒雪、风餐露宿、跋山涉水近10天，不断追寻，结果是只听猴啼叫而不见猴影来，只能无功而返。好在自然保护区的救护站有一只被救护的怒江金丝猴个体，这也是世界上唯一一只在人工环境下生活的怒江金丝猴。看到它，我激动万分，把自己身上带的财物都捐献出来交给了救护站，以深深地感谢此行此福！

China is one of the world's centers where primates were originated and evolved into different species, and the Hengduan Mountains in the SWR is the core region where primates in China are distributed. China is home to four kinds of golden monkeys (*Rhinopithecus*). Except for gray golden monkey (*Rhinopithecus brelichi*) that is typically found in the CCR (Central China Region), the SWR has all the other three species of golden monkey —

↶ 不好意思，我们刚认识·云南泸水
A newly-acquainted friend — Lushui, Yunnan

怒江金丝猴是直到2010年才被发现的金丝猴新种，仅分布在中缅边界的怒江和恩梅开江之间的狭小区域内，迫切需要中缅两国建立共同的跨界机制予以保护。
Nujiang golden monkey is a new species of golden monkey that was not discovered until 2010. Distributed only in a narrow niche within the catchments of the Nujiang River and Nmai Hka along the borderlines between China and Myanmar, this is a wildlife species that calls urgently for the two countries to set up a joint, cross-border mechanism for its protection.

↶ 我是老大·四川宝兴
I am the lord — Baoxing, Sichuan

中国4种金丝猴中分布最广、数量最多的是川金丝猴，除四川西部外，秦岭和神农架都是它们生活的地方。它是金丝猴中最早（1870年）被发现并命名的。
China is home to four kinds of golden monkeys, among which Sichuan golden monkey is most widely distributed and has the largest population. Besides the western Sichuan, they are also found to exist in the Qinling Mountains and Shennongjia Mountains. Among all the four species, this species was discovered and named the earliest (in 1870).

Sichuan golden monkey (*Rhinopithecus roxellana*), black golden monkey (*Rhinopithecus bieti*), and Nujiang golden monkey (*Rhinopithecus strykeri*). Golden monkeys are of medium size among primates, with the tail as long as the body. Golden or black gray are their main fur color. Their main feature is that their nostrils are almost parallel to their face, so they are also called snub-nosed monkeys. Researches suggest that reduced nasal bones help to decrease the resistance of breathing in the thin air, which is an adaptation to the hypoxia environment in the plateau.

Gibbons (Hylobatidae) are slightly smaller than golden snub-nosed monkeys and have no tails. They are named "long armed apes" in China after their short legs and long arms. Their palms are longer than the soles of their feet, and the knuckles of their hands are longer than those of their toes, which is why they mainly rely on long arms to move and leap between trees. The male gibbon is generally of black or dark brown color, with comparatively darker fur. But the female and young ones have brownish yellow or golden fur. There are eight species of gibbon in China, all of which are listed as Class I Key Protected Wild Animal of National Significance. Except for Hainan gibbon (*Nomascus hainanus*), eastern black crested gibbon (*Nomascus nasutus*) and white-cheeked gibbon (*Nomascus leucogenys*) that primarily live in the SCR (South China Region), the SWR is home to western white-browed gibbon (*Hoolock hoolock*), eastern white-browed gibbon (*Hoolock leuconedys*), western black crested gibbon (*Nomascus concolor*), white-handed gibbon (*Hylobates lar*) and Gaoligong white-browed gibbon (*Hoolock tianxing*). The scientific name of apes is Hominoidea due to their appearance that is very much similar to that of human beings. In terms of lineage, they are also closer to human beings. The most striking difference between apes and monkeys in appearance is that the former don't have tails and cheek pouches, whereas the latter do.

Leaf monkeys or langurs (*Presbytis*) mainly feed on leaves and buds and rely on the nutrients and water derived therefrom to sustain their lives, for which reason they are thus named. China is home to seven leaf monkey species. Except for the François's

萌人的叶猴·云南无量山
Beloved langur — Wuliang Mountains, Yunnan

菲氏叶猴是比较典型的南亚热带树栖叶猴。

Phayre's langur is a typical tropical arboreal langur inhabiting South Asia.

langur (*Trachypithecus francoisi*) that chiefly lives in the CCR, the Nepal gray langur (*Semnopithecus schistaceus*) that chiefly lives in the QTR and the white-headed black langur (*Trachypithecus poliocephalus* ssp. *leucocephalus*) that is mainly found in the SCR, all the other three species — Indo-Chinese gray langur (*Trachypithecus crepusculus*), Phayre's langur (*Trachypithecus phayrei*), capped langur (*Trachypithecus pileatus*) and Shortridge's langur (*Trachypithecus shortridgei*) — are distributed in the SWR.

The most commonly-seen monkey is the macaque (*Macaca*). The macaque is characterized by their short tails, stocky bodies, cheek pouches for food storage, and forelegs that are of roughly the same length as their hind legs. China has eight species of macaque, among which the SWR boasts six — rhesus macaque (*Macaca mulatta tcheliensis*), Tibetan macaque (*Macaca thibetana*), bear macaque (*Macaca arctoides*), Assamese macaque (*Macaca assamensis*), northern pigtail macaque (*Macaca leonina*) and white-cheeked macaque (*Macaca leucogenys*), the exceptions being formosan macaque (*Macaca cyclopis*) that mainly lives in the SCR region and Arunachal macaque (*Macaca munznla*) that mainly lives in the QTR.

Notably, the SWR is the endemic home to the Nujiang golden monkey (*Rhinopithecus strykeri*) discovered in 2010, the white-cheeked macaque (*Macaca leucogenys*), a new monkey species discovered and named by the Chinese scientists in 2015, and the Gaoligong white-browed gibbon (*Hoolock tianxing*), a new gibbon species discovered and named by Chinese scientists in 2017. Some scientists believe that more new species belonging to the primate family may be found in the region in the future. This region indeed deserves its reputation as the paradise of primates!

The primates live primarily on trees within forest ecosystems. They are mostly social animals, living and migrating in groups. Except for the golden monkeys and gibbons that live in families, most of the other monkeys live in communities headed by their respective monkey kings.

Extremely rare and precious, all primates in China are listed as Class I or Class II Key Protected Wild Animal of National Significance. Throughout my decades-long career in wildlife protection, I've seen most of them in the wild, except for Nujiang golden monkeys which has been most difficult to locate. In the deep Nujiang Canyon covered with exuberant forests, we travelled against extreme weather over hills and rivers for nearly 10 days with no sight of them but their sound only. Fortunately, I saw in the rescue station of a nature reserve a Nujiang golden monkey that had been rescued by the local rangers — the only one that survives under man-made environment. I was so excited upon the sight of this creature that I donated everything I had with me at that time to the rescue station, just to show how grateful I was for being blessed with such a precious chance to see it in person.

可爱的红嘴唇·云南白马雪山
Lovely red lips — Baimaxueshan, Yunnan

滇金丝猴有一张和人类一样的红嘴唇，十分可爱！它们只分布在云南西北部以及云南和西藏交界的高山针叶林中。图为滇金丝猴亲密的一家。

The Yunnan golden monkey (*Rhinopithecus bieti*) is born with red lips that are much similar to those of human beings, making it appear incredibly lovely and cute. This species of monkeys is found only in the northwestern Yunnan and the alpine coniferous forests grown along the border between Yunnan and Tibet. This photo shows the close family of Yunnan golden monkeys.

天行之猿·云南高黎贡山
Skywalker hoolock gibbon — Gaoligong Mountains, Yunnan

天行长臂猿是高黎贡白眉长臂猿的另外一个称呼，它是首个由我国科学家命名的长臂猿种，是全球最濒危的25种灵长类之一。"天行"来源于道家的"天行健，君子以自强不息"，命名者希望像道家倡导的那样遵循自然规律做谦谦君子。

The skywalker hoolock gibbon (*Hoolock tianxing*), another name of Gaoligong white-browed gibbon, is the first gibbon species that was named by Chinese scientists. The species is one of the world's 25 most endangered primates. The term *tianxing* (literally meaning the skywalker) is derived from the Taoism doctrine that prescribes "as heaven maitains vigor through movement, a gentleman should constantly strive for self-perfection". As implied, the scientists who named the animal this way wish that the natural ways as advocated in Taoism can be followed in our wildlife protection efforts.

午觉醒来·云南无量山
❶ Waking up from its midday nap — Wuliang Mountains, Yunnan

黑冠长臂猿各亚种中，海南长臂猿分布于海南岛霸王岭，东黑冠长臂猿分布于广西南部的中越交界线附近，西黑冠长臂猿分布于云南无量山、哀牢山一带。图中的西黑冠长臂猿在树上午休醒来，正伸懒腰呢。

Among the subspecies of black crested gibbon, the Hainan gibbon (*Nomascus hainanus*) is distributed in Bawangling on the Hainan Island; the eastern black crested gibbon (*Nomascus nasutus*) live primarily in the southern Guangxi that borders Vietnam; and the western black crested gibbon (*Nomascus concolor*) inhabits mainly in the Wuliang Mountains and the Ailao Mountains of Yunnan. The one in the photo, which belongs to the last subspecies, seems to have just waken up from its midday nap and is stretching itself out in a lazily manner.

母子俩·四川唐家河
❷ A mother monkey and her baby — Tangjiahe, Sichuan

藏酋猴是猕猴属中体型最大的一种，为中国特有种。其颜面部分仔猴为白色，长大后逐渐变红，老年后呈紫色和紫黑色。图中母子面部各色。

The short-tailed Tibetan macaque (*Macaca thibetana*), an endemic species to China, is a relatively large-sized species among the genus *Macaca*. The color in its face, which is white upon birth, changes gradually as the monkey grows, turning reddish at first and finally into purplish or dark brown in its senior years. The mother and her child in the photo are clearly distinguishable from the distinctive color in their faces.

路边客·云南保山
❸ Regular visitors to the highways — Baoshan, Yunnan

熊猴因体型肥壮，憨态似熊而得名，是胆子最大的猴之一。这是经常在翻越高黎贡山的公路边溜达的雄性熊猴。

Assamese macaque (*Macaca assamensis*, bear monkey in Chinese) is thus named because of its stout and clumsy manners much similar to that of bears. It is probably one of the boldest species among monkeys. This male Assamese macaque in the photo shows regularly up to take its leisurely strolls along the highway that zigzags through the Gaoligong Mountains.

打洞高手·云南思茅
Master in burrowing — Simao, Yunan

穿山甲是地栖性哺乳动物，擅长打洞，喜欢昼伏夜出。由于人为的大量猎捕，穿山甲已由常见物种变成了极度濒危物种。

The pangolin (*Manis pentadactyla*), a terrestrial mammal, is especially good at burrowing and prefers to hide during daytimes and comes out only at nights. As a consequence of exploitative hunting by human beings, this mammal has been reduced in number from a widely-seen species to one that is critically endangered (CR).

四川羚牛·四川唐家河
Sichuan takin — Tangjiahe, Sichuan

羚牛分为4个亚种，即贡山羚牛、不丹羚牛、四川羚牛和秦岭羚牛，在中国均有分布，由于产地不同，毛色由南向北逐渐变浅。图为四川羚牛。

The takin (*Budorcas taxicolor*) has four subspecies, including: Mishmi takin (*Budorcas taxicolor*), Bhutan takin (*Budorcas whitei*), Sichuan takin (*Budorcas tibetanus*) and Shensi takin (*Budorcas bedfordi*), all of which can be found in China. Due to different geographical areas where they typically inhabit, the color of their fur also changes gradually, that of those in south is lighter than those in the north. The one in the photo is a Sichuan takin.

会飞的松鼠·云南德宏
Flying squirrel — Dehong, Yunnan

红白鼯鼠属夜行性动物，其前后肢间有飞膜，借此能在树与树、树与地之间滑翔，也称"飞鼠"。

The red-and-white giant flying squirrel (*Petaurista alborufus*), a nocturnal animal, has flying membrane between its front and rear limbs, which allows it to glide from tree to tree or ground. For this reason, it also known as the flying squirrel.

多种多样的鸟类·西南区
Variegated birds — SWR

西南区的鸟种占全国鸟种的一半左右。在各种生境中，亚热带低山常绿阔叶中林鸟种类最为丰富，其次是暖温带中山针阔混交林，而针叶林、亚寒带高山灌丛草甸、河湖沼泽、村庄农田逐渐次之，干热河谷灌丛最为贫乏。图中鸟种依次为：蓝翅希鹛（图1）、大拟啄木鸟（图2）、白胸苦恶鸟（图3）、白胸翡翠（图4）、栗额斑翅鹛（图5）、红嘴鸥（图6）、血雀（图7）、紫水鸡（图8）、金背啄木鸟（图9）、黄颈凤鹛（图10）、银耳相思鸟（图11）。

The SWR accounts for about half of the country's total bird species. Among the various habitats, the subtropical low mountain evergreen broad-leaved forests tend to have the richest diversity of bird species, followed by the middle moutain coniferous and broadleaved mixed forest of warm temperate zone, which is in turn followed by the coniferous forests; the alpine shrub meadow of subfrigid zone; rivers, lakes and marshes; and finally villages and farmlands. Shrubs in dry and hot valleys tend to have the smallest diversity of birds. The birds in these photos are: *Siva cyanouroptera* (Photo 1), *Psilopogon virens* (Photo 2), *Amaurornis phoenicurus* (Photo 3), *Halcyon smyrnensis* (Photo 4), *Actinodura egertoni* (Photo 5), *Chroicocephalus ridibundus* (Photo 6), *Carpodacus sipahi* (Photo 7), *Porphyrio porphyrio* (Photo 8), *Dinopium javanense* (Photo 9), *Yuhina flavicollis* (Photo 10) *and Leiothrix argentauris* (Photo 11).

中国新鸟种·云南蒙自
A new bird species in China — Mengzi, Yunnan

钳嘴鹳为热带湿地的迁徙鸟，南亚、东南亚物种，2006年10月在云南大理首次记录后，现在中国多处被发现，分布范围也越来越广。有专家认为这是全球气候变暖造成的气候带北移所致。

The Asian openbill (*Anastomus oscitans*), a migratory bird that primarily inhabits wetlands of tropical regions, is a typical species of South and Southeast Asia. After being first discovered in October 2006 in Dali, Yunnan Province, it has now been found in increasingly more places in China. Some experts argue that it is caused by global warming and the northward movement of the climate zones arising therefrom.

越冬的黑颈鹤·
云南大山包
Black-necked crane
overwintering —
Dashanbao, Yunnan

云贵高原东部是黑颈鹤最佳的越冬地,海子(小湖泊)和庄稼地是它们休息和觅食的最佳场所,村民是它们不离不弃的最好伙伴。
The eastern Yun-Gui Plateau is the best place for the black-necked crane (*Grus nigricollis*) to overwinter. Haizi (small lakes) and farmlands make up ideal places for them to have a rest and forage. Local villagers have become their good friends that keep company for them all the time.

对"野生动物王国"的保护建议
Recommendations for the Conservation of the "Kingdom of Wildlife"

生态思考 / Ecological Reflection

西南区是世界34个生物多样性热点地区之一，也是中国野生动物最为丰富的地区，这里是中国的"野生动物王国"。

放在全球范围来讲，中国的整个西南区地理条件都异常独特，生物区系复杂而多样，从海拔几百米的低湿河谷一直到海拔7000米以上的极高山地，分布着从河谷季风雨林一直到高山苔原几乎所有的植被类型。这个位于世界温带区域物种最丰富的地区拥有大量独有的动植物物种。该地区的面积大约占中国陆地面积的10%左右，却拥有全国50%的鸟类和哺乳动物种类以及30%以上的高等植物。它不仅有大熊猫这样的古老区系成分，而且是中国灵长类的分布中心、中国雉类的分布中心，也是噪鹛类和鸦雀类等的分布中心及可能的起源地，被科学界誉为"世界物种的基因库"。

尽管我们在这块土地上已经建立了大批的自然保护区，在保护方面投入了很多的人力、物力和财力。但是，对于这块中国特有乃至世界难得的生物多样性富集地区，我们还有很长的路要走。以灵长类为例：2010年，发现了金丝猴新种——怒江金丝猴，尽管我们先后创造了首次拍到该金丝猴在野外生存的照片和视频影像，首次在分子水平上肯定了它在中国的存在，首次经救护获得可近距离观测研究的活体等多个人类认识怒江金丝猴的世界纪录，但十分遗憾的是最早发现它是在缅甸而不是在中国。2015年，中国科学家定名的新种——白颊猕猴，2017年中国科学家新命名的高黎贡白眉长臂猿都在这里。有科学家预测，这个地区可能还存在着灵长类新种而未被人类发现。

这里的野生动物物种丰富但种群数量小，分布区狭窄，所建自然保护区的面积都不太大，很多自然保护区呈狭长的不稳定状态，这些栖息地特别脆弱，一旦遭受割裂或破坏，物种就会迅速陷入濒危以致走向灭亡且难以挽回。与其他区域相比，这里野生动物保护的客观条件苛刻、保护难度大。我们尤其需要提高对这块区域生物多样性保护的重视,《生物多样性公约》的"爱知目标"要求到2020年至少有17%的陆地及内陆水域应该建立自然保护地以有效保护生物多样性,这是一个平均值。如果这块地方我们做不到"自然需要一半",最少1/4~1/3应该有吧?"大熊猫国家公园"的四川园区占地面积达2万平方千米,我们能否也像大熊猫保护一样,投入建设高黎贡山、灵长类、雉鸡类国家公园,让这些特殊生态系统和伞护种像大熊猫一样庇护这块中国乃至全球不可多得的"野生动物王国"?

The South-west Region (SWR), which boasts the greatest wildlife biodiversity in China, is one of the 34 biodiversity hotspots of the world and hailed as the "Kingdom of Wildlife".

From global perspective, the SWR in China has incredibly distinctive geographical conditions, and complex and diverse biota. From wet valleys that are just a few hundred metres above sea level to extremely high mountains that are over 7,000 metres in altitude, almost all vegetation types ranging from valley monsoon rainforests to alpine tundra can be found here. The temperate SWR has the greatest biodiversity among its worldwide counterparts, with a fairly large proportion of the wildlife species being endemic to this region. Though merely taking up around 10 percent of China's land area, the SWR is home to 50 percent of birds and mammal species, and over 30 percent of higher plants found in the country. It is not only home to some ancient species like giant pandas, but also a center where primates, pheasants are mostly densely distributed. It is highly likely that this is also the originating place as well as distribution center of *Garrulax* and *Paradoxornis*. For the above reasons, the SWR is regarded by the community of scientists as the "gene bank of life species on the Earth".

We have established a large number of nature reserves in the SWR and invested a lot of manpower, material resources and financial resources. However, there is still a long way for us to go in fulfilling our responsibilities for protecting this biodiversity-rich region that is of critical importance to both China and the world. Taking the primates as examples: Nujiang golden monkey (*Rhinopithecus strykeri*) is a new species of golden monkeys discovered in this region in 2010. We have taken pictures and videos of this species; we have been able to watch from short distance at an individual that are kept at the rescue center; we have proven the existence of this peculiar species in China at molecular level. However, in spite of all these world records we have made in our understanding about this animal, it remains a great pity that the earliest discovery of this monkey was made in Myanmar rather than in China. Both white-cheeked macaque (*Macaca leucogenys*) and Gaoligong white-browed gibbon (*Hoolock tianxing*) ,which were discovered and named by Chinese scientists in 2015 and 2017 respectively, live in the region. Some scientists believe that more new species belonging to the primate family may be found in the region in the future.

But given that the populations of wildlife species here are typically very small and their habitats are often narrow, nature reserves set up for their protection are often not large in scale, forming narrow strips of fragile niches that, once isolated or destroyed, will lead to irreversible decline, or even extinction of the target species. Compared with other regions, the SWR is faced with more challenging situations in wildlife conservation. We should attach even greater importance to biodiversity conservation in the SWR. According to the Aichi Biodiversity Targets adopted by the United Nations *Convention on Biological Diversity*, nature reserves should be set up on at least 17 percent of the land and inland waters in the world so that global biodiversity is put under effective protection. This percentage is only the mean figure. But for such a precious region like the SWR, even if the 50%-goal can't be reached, we should strive for the goal of setting at least 1/4-1/3 of this region aside for nature reserves. The size of the Giant Panda National Park in Sichuan amounts to 20,000 km^2. Is it possible that we also set up similar national parks for Gaoligong Mountains, primates and pheasants, so that these special ecosystems, together with the precious umbrella wildlife species thereof, can be safely protected, just as the pandas are, in this Kingdom of Wildlife which is invaluable not only for China but for the world at large?

树上的小熊猫·四川岷山
A red panda on the tree — Minshan Mountains, Sichuan

小熊猫比大熊猫早发现半个世纪,它并不是小的大熊猫,小熊猫为小熊猫科、大熊猫为大熊猫科动物。其生活区域常常和大熊猫重叠,同样喜食箭竹的竹笋、嫩枝和竹叶,也食各种野果、树叶、苔藓等,喜欢树栖。

The red panda (*Ailurus fulgens*) was discovered half a century earlier than the giant panda. Instead of being, as implied by its name in Chinese, a type of giant pandas that are comparatively smaller in sizes, red pandas belong to a totally different family — Ailurinae, whereas giant pandas belong to the family of Ailuropodidae. The habitats of the two animals overlap to a large extent. Like the giant pandas, the arboreal red pandas are also fond of eating tender shoots of *Fargesia spathacea* Franch, burgeon and bamboo leaves, as well as various wild fruits, leaves and mosses.

入湖河流及沙洲·江西鄱阳湖
Streams emptying into the Poyang Lake and the sandy islets formed thereof — Poyang Lake, Jiangxi

鄱阳湖是中国最大的淡水湖、长江最大的通江湖泊，这里孕育了无数的陆生、水生生物，是中国乃至世界最著名的湿地生态系统之一。
The Poyang Lake, the largest fresh water lake in China and the greatest lake connected with the Yangtze River, is home to numerous terrestrial and aquatic species and considered as one of the most famous wetland ecosystems in China and even the world.

华中区

Central China Region

秦岭深处·陕西长青
In the depths of the Qinling Mountains — Changqing, Shaanxi

高耸的秦岭挡住了北方来的寒流，庇护了华中、华南温润的气候，为中国南方万物生长提供了优良的条件。
The lofty Qinling Mountains have blocked the cold current from the north, thus shielding the mild and humid climate in Central and South China and enabling wildlife in the southern China to grow vigorously.

地带性植被·浙江钱江源
Zonal vegetation — Qianjiangyuan, Zhejiang

在世界同纬度区域，中国的亚热带比较宽大广阔，这里是亚热带常绿阔叶林的主要分布区。
Compared with other regions in the world that are situated at similar latitudes, China has a relatively wider scope of subtropical climate zone, thus making this place a key area where subtropical evergreen broad-leaved forests densely concentrate.

长江三峡·湖北宜昌
The Yangtze River Gorges — Yichang, Hubei

长江，这条中国第一大河，是中华民族的母亲河，其宽广的流域养育了无数的野生动植物。
The Yangtze River, the largest river in China as well as the mother river of the Chinese nation, supports numerous wild animals and plants with its wide drainage area.

华中区

北界自秦岭南坡、伏牛山、大别山一线向东，大致沿淮河流域南部到长江以北的通扬运河一线，南界自福州向西沿戴云山经南岭南侧和广西瑶山，西与西南区为邻，包括秦岭南坡、汉中盆地、川西平原、贵州高原、东部丘陵、长江中下游平原及东海等。本区属世界动物地理区划中东洋界的中国东部分，气候区为北亚热带和中亚热带。本区动物区系主要由东洋型和南中国型组成，很多物种与华南区共有。

秦岭是中国的"中央公园"，不仅是中国温带和亚热带的分界线，也是中国南北方的分界线。秦岭野生动物物种非常丰富，以著名的"秦岭四宝"——大熊猫、川金丝猴、秦岭羚牛、朱鹮为代表的野生动物多种多样。经过这些年来的努力，秦岭建立了以佛坪、太白山、周至、长青、洋县等为代表的自然保护区集群，较好地保护了生存在这片区域的野生动物。长江是中国重要的生态宝库，是世界上水生生物多样性最为丰富的河流之一，分布有水生生物4000多种。由于长江运输航线的日益繁忙，水利工程的实施，工业及民生等大量污水的排放，长江江豚、中华鲟、长江鲟等为代表的水生野生动物的生存堪忧。在长江江岸河滩、通江湖泊及长江口海岸滩涂湿地上，包括鄱阳湖、洞庭湖等众多湖泊中，大量迁徙的鸟、扬子鳄及野放的麋鹿等种群数量得到了较大的保护和发展。总之，该区域野生动物物种生存状况有喜有忧。

2016年，中央批准了《长江经济带发展规划纲要》，规划的要求是，实施长江经济带发展战略必须从中华民族长远利益考虑，把修复长江生态环境摆在压倒性位置。2018年，国务院又发出《关于长江水生生物保护工作的意见》，拯救长江水生野生动物的行动已经全面铺开，长江经济带从"大开发"转到"大保护"的战略方针指导下，其生态环境得到了保护和恢复。建议在现有神农架、武夷山、南山、钱江源等国家公园试点建设的基础上，积极推进秦岭、长江（湖泊、河口）等国家公园的建设，以国家公园建设为主体积极推进自然保护地体系建设，在"自然保护地分级分类分区"管理上狠下功夫，加大野生动物的保护力度，使整个华中区的生物多样性保护和恢复更上一层楼。

The northern boundary of the Central China Region (CCR) extends from the southern slope of Qinling Mountains, the Funiu Mountains and the Dabie Mountains eastward to the Tongyang Canal that lies roughly in the belt between the southern part of the Huai River Basin and the north of the Yangtze River. The southern boundary of CCR starts from Fuzhou and extends in westward direction along the Daiyun Mountain, the southern slope of the Nanling Mountains and the Yao Mountain in Guangxi. The west of CCR is adjacent to the South-west Region (SWR). The CCR covers within its scope such places as the southern slope of the Qinling Mountains, the Hanzhong Basin, the western Sichuan Plain, the Guizhou Plateau, the eastern hills, the plain along the middle and lower reaches of the Yangtze River and the East China Sea. In global zoogeographic zoning, this region belongs to the eastern part of the Oriental Realm within China, and falls into the northern subtropical and the central subtropical climate zones. The prevailing fauna species in this region are mainly of the Oriental and the South China types, many of which are the same as those inhabiting the South China Region (SCR).

The Qinling Mountains, deemed as the Central Park of China, serves as both the borderline between the temperate and subtropical zones in the country and the borderline between the northern and southern parts of China. Wildlife inhabiting in Qinling Mountains is extraordinarily diversified, represented by the "Four Treasures of the Qinling Mountains": the giant panda (*Ailuropoda melanoleuca*), the Sichuan golden monkey (*Rhinopithecus roxellana*), the golden takin (*Budorcas tibetanus*) and the crested ibis (*Nipponia nippon*). After years of efforts in nature conservation, a sophisticated network of nature reserves has come into being in and along the Qinling Mountains — represented by Foping, Mount Taibai, Zhouzhi, Changqing and Yangxian, ensuring better protection of wildlife inhabiting in this region. The Yangtze River, an important ecological treasure-house in which over 4,000 aquatic species live, ranks among the most aquatic biodiversity-rich rivers in the world. On the one hand, as a consequence of increasingly busier ship routes operating in the Yangtze River, the construction of irrigation works, as well as of industrial and domestic pollutants discharged into the river, the Yangtze River dolphin (*Neophocaena asiaeorientalis asiaeorientalis*), the Chinese sturgeon (*Acipenser sinensis*), the Yangtze sturgeon (*Acipenser dabryanus*) and many other representative aquatic wildlife species endemic to the river are faced with worrisome conditions for survival. On the other hand, on the beaches of the Yangtze River bank, in lakes that feed into the Yangtze River, and on the tidal flat wetlands in Yangtze River delta, such as the Poyang and the Dongting lakes, great efforts have been made in the protection of migratory birds, the Chinese alligator (*Alligator sinensis*) and the reintroduced David's deer (*Elaphurus davidianus*), leading in turn to obvious recovery and even great growth in the population of the protected wildlife species. In short, the surviving condition for wildlife species in this region is both promising and worrisome.

In 2016, the Central Government Approved the *Outline of the Yangtze River Economic Belt Development Plan*, which requires that in the implementation of the development strategy of the Yangtze Economic Belt, long-term interests of the Chinese nation must be taken into consideration and the restoration of the ecology in the Yangtze River must be placed in an overwhelmingly important position. In 2018, the State Council issued *the Opinions on the Protection of Aquatic Life in the Yangtze River*, marking the beginning of an extensive campaign to rescue aquatic wildlife living in the river. Under the guidance of strategies that urge a fundamental transformation from "development-centered approaches" to "conservation-centered approaches" in the Yangtze River Economic Belt, inspiring achievements have been made in ecological restoration efforts. It is suggested that, in addition to the pilot national parks that have been established in Shennongjia, Wuyi Mountains, Nanshan Mountain and Qianjiangyuan, more such national parks should be set up in Qinling Mountains, Yangtze River (including corresponding lakes and estuaries) and other places. We will vigorously enhance the development of a natural protected area system highlighting the central role that national parks play, stress on approaches that advocate "the adoption of tiered and diversified management to various protected areas", and further strengthen our efforts for wildlife protection. In doing so, further improvement will be achieved in this region in the preservation and restoration of biodiversity.

↑ 水中嬉戏·湖北神农架
Playing in the water — Shennongjia, Hubei

在长江一些支流里活跃着主要以鱼类为食物的水獭。水獭是优良水环境的指示物种。
Living in some tributaries of the Yangtze River are otters (*Lutra lutra*), which feed mainly on fish and are considered as an indicator species for sound water environment.

国际合作共同保护"东方宝石"的故事
A Story of International Cooperation to Protect the "Oriental Gem"

雪中朱鹮·陕西长青
A crested ibis in snow — Changqing, Shaanxi

在冰冻雪封的日子里，秦岭南坡的河道中，朱鹮在雪上飞来跳去寻觅食物。
A crested ibis (*Nipponia nippon*) is foraging carefree in the snow-covered river course sitting at the foot of the southern slope of the Qinling Mountains.

朱鹮系东亚特有种，曾广泛分布于中国东部、日本、俄罗斯和朝鲜半岛等地，栖息于湿地的疏林地带，在高大的树木上休息和筑巢，在附近的溪流、沼泽及稻田里觅食鱼、蛙、泥鳅及螺等。

20世纪中叶以来，由于大肆猎捕、湿地减少和使用农药等因素，朱鹮生存空间不断被破坏，种群数量急剧下降。俄罗斯、朝鲜半岛的种群先后消亡，在中国最后一次见到朱鹮是在1964年，日本到1980年仅剩5只，之后日本将这野外的5只全部捕获用来人工饲养（至1983年陆续死完，至此，中国之外再无朱鹮）。1981年5月，中国科学家在陕西洋县的山林中发现7只朱鹮。中国采取了强化就地保护为主、人工易地保护为辅的正确策略，无论是野外的还是人工饲养的朱鹮保护效果都很明显，种群数量得到了极大扩展。

朱鹮有着鸟中"东方宝石"之称，洁白的羽毛，艳红的头冠和黑色的长嘴，加上细长的双脚，非常美丽，历来被日本皇室奉为圣鸟。朱鹮的拉丁学名"*Nipponia nippon*"直译为"日本的日本"，以国名来命名鸟名，足见朱鹮对于日本的重大意义。

为了共同保护好朱鹮这个珍贵物种，在日方的积极推动下，中国和日本两国于1985年签订了《中日共同保护朱鹮计划》。日本给中国以设备、资金方面的支持，并从中国借3只朱鹮用于繁育，但至1995年，这三只朱鹮只剩下了"阿金"一只。

1998年，国家主席江泽民向日本赠送一对朱鹮。2000年，朱镕基总理又将朱鹮"美美"（雌）送给了日本用于朱鹮配对繁殖。中国不仅将朱鹮种源基因送给了日本，而且在中国培训了日本的技术人员，中方的专家又到日本佐渡岛长期亲临指导。这样，日本的朱鹮种群终于渡过难关，发展了起来。我当时就负责具体操作此事，整个过程至今历历在目。

目前，中国的朱鹮种群已经发展到了5000多只，日本的种群也发展到了500只左右，中国赠送给韩国的朱鹮也由4只繁衍到了300多只。现在，朱鹮已经能够自由自在地飞翔在中华大地、日本佐渡岛、韩国昌宁的上空，这不仅仅是中国野生动物保护成功的典型案例，更是一曲由中国主导的、国际合作共同拯救和保护"东方宝石"朱鹮的凯歌。

The crested ibis (*Nipponia nippon*), an endemic species of East Asia, used to be widely distributed over eastern China, Japan, Russia and Korean Peninsula. Inhabiting the open forest zone in wetlands, it typically perches and builds nests on tall trees, and forages for fish, frogs, loaches and snails in nearby streams, swamps and paddy fields.

Owing to such factors as excessive hunting, reduction of wetland and the use of pesticide since the middle of the 20th century, living space of the crested ibis has been destroyed continuously and its population has declined sharply. Its populations in Russia and Korean Peninsula disappeared successively. In China, it was in 1964 that last sighting of the crested ibis in the wild was reported. By 1980, the number of crested ibises surviving in the wild in Japan had dropped to five, all of which were later captured and kept in captivity (and died out one after another by 1983. By that time, the crested ibis had all gone extinct except for in China). In May 1981, Chinese scientists found seven crested ibises in the mountain of Yangxian County, Shaanxi Province. Thanks to the correct protective strategy that China has adopted, i.e., supplementing the central role of *in-situ* conservation measures with artificial *ex-situ* conservation endeavors, the population of crested ibises both in captivity and in the wild has been greatly boosted.

The crested ibis is known as the "oriental gem" of birds. With white feathers, red crown, black beak, and slender feet, it is such a beautiful species that it has always been regarded as a sacred bird by the Japanese royal family. *Nipponia nippon*, Latin name of the crested ibis, is translated literally as "Japan of Japan". The fact that the Japanese would choose to name a bird after the very name of the country is a telling evidence about the significant meaning that the bird has for the people.

In order to protect the crested ibis with joint hands, and under the active advocating of Japan, China and Japan signed the *China-Japan Joint Protection Plan of Crested Ibis* in 1985, according to which Japan provided China with equipment and funds, and three crested ibises were loaned from China for the purpose of propagation in Japan. By 1995, however, only one among the three birds, A Jin, was still living.

In 1998, President Jiang Zemin presented Japan with a pair of crested ibises. In 2000, Prime Minister Zhu Rongji presented another female crested ibis, named Mei Mei, to Japan for mating and breeding. China not only sent genes of the crested ibis to Japan, but also trained Japanese technicians in China. In addition, Chinese experts also went regularly to Sado Island on long-term basis for on-site training. Therefore, Japan has seen a gradual increase in the population of crested ibises. I was in charge of the protection plan at that time, and the whole process is still vivid in my mind.

Currently, the population of crested ibises in China has gone beyond 5,000, and that in Japan stands at around 500. Moreover, the four crested ibises that China presented to South Korea as gifts have also grown to more than 300. The crested ibis is now able to fly freely over China, Sado Island of Japan and Changnyeong of South Korea. This is not just a telling example that showcases the successful wildlife conservation in China, but a song of triumph that celebrates the joint efforts pioneered by China for international cooperation in rescuing and protecting the crested ibis — the "oriental gem".

 希望中成长·陕西洋县
Growing up with hope — Yangxian, Shaanxi

觅食归来的朱鹮妈妈正在给巢中渐渐长大的雏鸟们喂食。在父母的精心照料下，孩子们在希望中茁壮地成长。

The mother crested ibis (*Nipponia nippon*) returning from their foraging is feeding her young chicks in the nest. Thanks to their caring parents, the chicks are thriving with a bright future.

中华"龙"的文化溯源
Tracing the Cultural Root of the Chinese "Dragon"

大快朵颐·安徽宣城
A gratifying feast — Xuancheng, Anhui

扬子鳄作为肉食性的爬行动物，可以狼吞虎咽大快朵颐，也可以忍饥挨饿长期不食，这大概是它能够作为"活化石"生存下来的秘诀吧。
As a carnivorous reptile, the Chinese alligator (*Alligator sinensis*) is capable of gobbling down vast amount of food at one time, and putting up with long periods of starvation when no food is available, which probably explains why it has survived the long history of the planet as a "living fossil".

扬子鳄古称鼍(tuó)，中国特有种，在经历了最近一次的地球生物大绝灭之后残存了下来。在扬子鳄身上，至今还可以找到早先恐龙类爬行动物的许多特征，因此人称"活化石"。扬子鳄以鱼、虾、软体动物及昆虫为食，处于食物链顶级，对于维护当地生态系统平衡有着重要的作用。

历史上，扬子鳄曾经广布于黄河和长江的中下游及浙江南部的湿地中，由于人类长期以来大量侵占湿地及对其捕杀，扬子鳄的生存空间不断被压缩，至20世纪70年代，野生扬子鳄数量已经锐减至不到300条。1975年，中国在安徽宣城建立了世界上唯一的扬子鳄自然保护区，1980年，人工繁殖扬子鳄成功，产下了第一批幼鳄。至今，扬子鳄野生及人工繁育的种群数量已经分别达到200多只和20000多只（人工饲养后放归野外的扬子鳄已成功实现自然繁殖）。扬子鳄的保护是中国野生动物保护工作中的又一成功案例。

中华民族是"龙"的传人，尽管现在"龙"形象已经是多种动物的复合体且神圣化了，但在"龙"的起源说中最早具象的蛇和鳄中，我更相信是扬子鳄的说法。黄河、长江中下游是中华农耕文化的发源地，先民能够广泛接触并就近观察到扬子鳄的行为，扬子鳄的水陆两栖生活习性，四肢、长尾和鳞甲，身体低频震动激起水花和嘴巴能够喷出雾气的生物学特性以及常在昏夜或雨前吼鸣的行为，最有可能成为先民农耕时祈祷老天下雨的愿望寄托载体。"龙"多与水相关联，上天入水、腾云驾雾、祈求雨水都与农耕文化紧密相关，而与蛇无关。"龙"无蛇的信子却有鳄的四肢，更是鳄与蛇生物形象特性的重要区别之一。再加上扬子鳄凶恶的外表和肉食者的特性让人惧怕，这样又尊恐又寄托的心理必将使人把它的形象神圣化，于是龙图腾的雏形就产生了。

中华文化早在甲骨文中就有关于鼍的记载，春秋时代的《诗经》中也有"鼍鼓蓬蓬"的诗句，鼍叫起来像敲鼓一样发出的声响，也许就是"龙"的发音。后来演化的"龙"其实是鸟、兽、鱼的复合动物，有鳄脚、马鬃、鹿角、虎眼、狮鼻、牛耳、鹰爪、鱼鳞之说。龙图腾和十二生肖（包括龙属）一样，是华夏先民对于野生动物与人紧密关系的认识和升华，"自然动物"变成了"人文动物"，成为了具有特定意义的文化符号。

The Chinese alligator (*Alligator sinensis*), known in ancient China as Tuo, is an endemic species of the country that had survived the last mass extinction of life on earth. Since the Chinese alligator still possesses many features that used to be characteristic of dinosaur and other early reptiles, it is hailed as "living fossils". Standing on the top of food chains, Chinese alligators feed on fish, shrimp, mollusks and insects, playing an important role in maintaining the balance of local ecosystem.

According to historical records, Chinese alligator used to be widely distributed in the middle and lower reaches of the Yellow River and the Yangtze River, as well as in the wetlands of the southern Zhejiang Province. As a consequence of human encroachment on wetlands and excessive killing of Chinese alligators over the past, the living space of Chinese alligators has been constantly narrowed. By the 1970s, the wild population of Chinese alligators had plummeted to less than 300. In 1975, China established the world's only nature reserve of Chinese alligators in Xuancheng, Anhui Province, where the first group of artificially bred Chinese alligators were born later on in 1980. So far, the wild and captive populations of Chinese alligators have increased to over 200 and 20,000 respectively (Chinese alligators that are bred in captivity, following their reintroduction to the wild, have already given birth to young alligators through natural propagation successfully). The conservation of Chinese alligators makes for another successful example of wildlife conservation in China.

Chinese people take pride in being the descendants of "dragon". Though the Chinese dragon that we take as our national totem is an imaginary sacred animal that blends the features of many different animals, I prefer to believe that it had its origin in the Chinese alligator rather than the snake which were often regarded as the earliest concrete symbol of the Chinese dragon in traditional folktales. Chinese farming culture emerged first on the Yellow River and the Yangtze River plains, where our ancestors could make use of the favorable conditions to closely contact with Chinese alligators and observe their behavior. Considering its biological properties such as living both on land and in waters, its limbs, long tail and scales, its low-frequency body vibrations which can cause water spray and that the mist would come out of its mouth, as well as the fact that it often roars in the dark or before rain, I believe that the Chinese alligators is the most likely divine beast where our ancestors can pin their hopes on for rain that are much needed for bumpy farming. "Dragon" is often closely associated with water. For instance, idioms describing dragon that the Chinese people are most familiar with "going to heaven and diving into the water", "riding clouds and fog" and "praying for rain", are all closely related with farming culture, but have little to do with snakes. "Dragons" do not have tongues like that of snakes, but share much similarity with crocodiles in their scale-covered limbs, a distinctive difference between snakes and crocodiles. Besides, combined with Chinese alligator's menacing appearance and carnivore-like nature, the reverence, fear and emotional attachment people hold in their mind will inevitably lead to the consecration of its image, hence coming into being the primitive form of culture that takes the dragon as the totem.

In Chinese civilization, Tuo was recorded as early as in inscriptions on bones or tortoise shells of the Shang Dynasty (16th–11th BC). According to the *Book of Songs*, the earliest collection of poems written during the Spring and Autumn Period (ca.771–476/403 BC), there is a line "Drums made from the skin of Tuo resonate loud". The sound that Tuo gives out is similar to the sound one hears when drums are beaten, which presumably is the same with the sound that "dragons" emit. The "dragon" evolved later into an image that is the mixture of bird, beast and fish, bearing the feet of crocodile, the hair of horse, the antlers of deer, the eyes of tiger, the nose of lion, the ears of ox, the claws of eagle, and the scales of fish. The Dragon Totem, like the Chinese Zodiac, is our Chinese ancestors' understanding and sublimation of the close relationship between wild animals and human beings. As a result, the "natural animals" becomes the "human animals" loaded with significant cultural connotations.

⬆ 水中游"龙"·安徽郎溪
Swimming "dragon" in the water — Langxi, Anhui

目前，扬子鳄只分布在中国长江下游较小范围内的湖泊、水塘和沼泽中。据考证，野生扬子鳄原来生存在长江中下游的广大地区，龙图腾的产生与中华先民的稻作文明息息相关。

At present, the Chinese alligator (*Alligator sinensis*) is only distributed in a relatively small range of lakes, ponds and marshes in the lower reaches of the Yangtze River. Evidence shows that wild Chinese alligators had previously inhabited the vast areas on the Yangtze Plain and that traditional Chinese culture that takes the dragon as its totem is deeply rooted in the rice-cultivating agricultural activities that our ancestors were engaged in.

华南虎回归野外的希望在哪里？
Wherein Does the Hope for Reintroducing South China Tiger to the Wild Lie?

何时回归野外？·
上海动物园
When will it be reintroduced to the wild? — Shanghai Zoo

华南虎在中国野外已经功能性灭绝。上海动物园自1958年开始饲养并展出华南虎，1959年首次繁殖成功，至今已先后共繁殖华南虎100多只，成为中国动物园中华南虎的最大种群。华南虎回归野外，既是我们的最大期望，也是我们面临的严峻挑战。

The South China Tiger (*Panthera tigris amoyensis*) has become functionally extinct in the wild in China. The Shanghai Zoo started its effort in breeding the tiger in captivity and displaying the tigers since 1958. Since the first successful captive breeding of the tiger in 1959, more than 100 south China tigers have been successively bred here, making the biggest population among those in Zoos. Reintroducing the South China tiger to the wild is our best wish. At the same time, it is our serious challenge.

华南虎是唯一仅分布在中国的虎，因此又称"中国虎"。研究认为，华南虎的特征最接近老虎的直系祖先——中华古猫，全世界的所有老虎亚种均来源于华南虎。历史上，华南虎曾广泛分布于华中、华南、西南的广阔地区。华南虎是典型的山地林栖动物，主要生活在中国南方亚热带的常绿阔叶林、常绿阔叶与落叶阔叶混交林和针阔混交林中，捕食对象包括野猪、鹿类等有蹄类动物。由于现代人口的增加及长期将老虎作为害兽的"打虎"，华南虎的栖息环境急剧变小并呈岛屿化，再加上华南虎食物链上的物种及其数量大量减少的原因，华南虎在野外就很难生存了。2000—2001年，国家林业局和世界野生生物基金会进行了全国野生华南虎及其栖息地的调查，那时我在国家林业局野生动植物保护司负责协调全国调查的成果，记得当时广东车八岭、福建梅花山、湖南莽山、江西宜黄和浙江庆元都报道过发现华南虎踪迹（并没有陕西），浙江当时甚至还召开过新闻发布会。但经仔细落实，这些"发现"主要都是老百姓的目击，或者发现疑似老虎脚印、粪便、毛发挂爪等。本次调查之后，国外一些学者认为野生华南虎已经灭绝，而国内一些学者并没有放弃搜索野生华南虎的一丝希望。

由于一个物种的繁衍需要自然种群的存在，自然种群还必须维持一定的个体数目并保持基因达到足够的杂合水平，种群才不会因为近亲繁殖而发生近交衰退。原华南虎分布的地区已经多年不见华南虎的踪迹，即便偶然发现一两只华南虎个体，华南虎的种群在中国野外已经不可能存在的严峻事实已经摆在我们面前。全国野生华南虎及其栖息地调查后的第七年（2008年），著名的"周老虎"事件发生，随后国家林业局派出的专家组调查结果认定，陕西镇坪县并没有野生华南虎的存在。

但是，华南虎的人工繁殖种群还在延续。多年来，全国共有16家动物园、野生动物园和华南虎拯救基地饲养过华南虎，包括上海、贵阳、苏州等地动物园，福建梅花山、广州长隆等地也卓有成效地进行了人工饲养，全国前后圈养的华南虎大概在200只左右。南非成功地使几个濒临灭绝的物种种群在野外扩大到可以繁衍下去的数量，受其启发，全莉女士夫妇牵头成立了"拯救中国虎国际基金会"，于2003年9月在南非开始了华南虎的野外驯养项目，其结果喜人，第三代华南虎幼崽在南非老虎谷保护区出生，使严重濒临灭绝的华南虎在老虎谷保护区的数量从7只发展到了15只以上。但非常遗憾的是，由于某些原因，南非野外训练的华南虎至今没能回归中国。

中国国内目前还没有经过野外生活驯化成功的华南虎，虎的野外繁衍需要尽可能大的领地及足够完善的食物链。华南虎回归野外的希望在哪里？

The South China tiger (*Panthera tigris amoyensis*) is the only tiger species endemic to China, hence also named "Chinese Tiger". Researches show that the characteristics of the South China tiger are the closest to those of its immediate ancestor — the ancient Chinese cat, and that all subspecies within the tiger family in the world are derived from the South China tiger. According to historical records, the South China tiger has been widely distributed in the vast areas of Central China, South China and Southwest China. The South China tiger is a typical arboreal animal that primarily ranges in the evergreen broad-leaved forests, deciduous broad-leaved and evergreen broad-leaved mixed forests as well as coniferous and broad-leaved mixed forests in subtropical climate zone in the southern China. Its preys include wild boars, deer and other ungulates. As a result of both population growth and the fact that tigers were for a fairly long period of time taken as a hazard for human and therefore a target of hunting, the habitats of the South China tiger are becoming smaller and increasingly fragmented. Besides, as the variety and number of animals that the South China tiger prey on are getting even scarcer, it is become all but impossible for South China tigers to survive in the wild. In 2000-2001, the State Forestry Administration (SFA) and the World Wide Fund for Nature (WWF) conducted a nationwide survey on the South China tiger and its habitats. At that time, I worked in the Department of Wildlife Protection of the SFA and was in charge of summing up the findings derived from the survey. As far as I remembered, sightings of South China tigers were reported in Chebaling in Guangdong, Meihuashan Mountain in Fujian, Mangshan Mountain in Hunan, Yihuang in Jiangxi and Qingyuan in Zhejiang (but not in Shaanxi), and a press conference was even specially held in Zhejiang for the sighting. However, careful investigation showed that all these alleged sightings didn't have solid basis, most of the witnesses were local people and in most cases only backed by highly suspectable evidence like tiger footprints, feces, hair, and so on. Some foreign scholars concluded that the wild South China tiger had already gone extinct, while some domestic scholars remained hopeful and didn't give up their searching for the beast.

The reproduction of a species is possible only when a sufficiently-large and viable community of such species exists in the wild so as to prevent inbreeding depression. No trace of tigers was found many years ago in the area where the South China tiger were originally distributed. Even if one or two individual tigers have indeed occasionally been spotted, a grim fact confronting us is that it is just unlikely that any viable community of wild South China tigers still exists in China. Seven years later on (in 2008) since the last nationwide survey of wild South China tigers and their habitats, a notorious event sprang up in China, in which a farmer surnamed Zhou claimed that he had found a tiger with evidence of fake photographs. After a thorough investigation, an expert panel from SFA concluded that there was no wild South China tiger living in Zhenping County in Shaanxi Province.

Nonetheless, captive-bred population of South China tigers still exist in China. Over the past decades, South China tigers have been raised in 16 zoos, wildlife parks and South China tiger rescue bases across the country. In such zoos like Shanghai, Guiyang and Suzhou, as well as in wildlife parks like Meihuashan Mountain of Fujian and Chimelong of Guangzhou provinces, artificial breeding of tigers has been successfully carried out. In total, about 200 South China tigers have been raised successively in captivity around the country. Inspired by South Africa's success in expanding several endangered species into viable populations in the wild, Ms. Quan and her husband established the Save China's Tigers Fund and launched the Wild Training Project for South China Tigers in September 2003 in South Africa that have yielded promising results. The third generation cubs of South China tigers were born in the South Africa's Tiger Valley Reserve, increasing the number of that

critically endangered species in that reserve from seven to more than 15. But unfortunately, the South China tiger trained in the wild in South Africa has not yet been able to return back to China for some reasons.

So far, there is not yet any successful cases in which captive-bred South China tiger has been trained to survive in the wild within China. Given that a viable community of tigers in the wild must be supported by a relatively large territory and a relatively complete food chain, wherein does the hope for reintroducing the South China tiger to the wild lie?

好奇的黔金丝猴・贵州梵净山
The curious gray golden monkey — Mount Fanjing, Guizhou

黔金丝猴活动的海拔高度比川金丝猴、滇金丝猴、怒江金丝猴都要低，活动的区域也最狭窄，仅分布于中国贵州省境内武陵山脉的梵净山自然保护区中。

The altitude where the gray golden monkey (*Rhinopithecus brelichi*) lives is lower than that of Sichuan golden monkey (*Rhinopithecus roxellana*), black golden monkey (*Rhinopithecus bieti*) and Nujiang golden monkey (*Rhinopithecus strykeri*). Moreover, this species also has the narrowest niche, distributed only in areas around the Mount Fanjing Nature Reserve located in Wuling Mountains of Guizhou Province.

① 正在下蛋的花龟・浙江金华
The striped-neck turtle laying its eggs — Jinhua, Zhejiang

花龟常生活于低海拔的水域，为高度水栖的淡水龟，杂食性，下蛋不易见。

Living in low-altitude waters, the striped-neck turtle (*Mauremys sinensis*) is a highly-aquatic turtle living in freshwater and is omnivorous in food. It is a rare event to see them laying eggs.

② 溪水里的大鲵・江西靖安
Chinese giant salamander in the stream — Jing'an, Jiangxi

大鲵是世界上现存最大的也是最珍贵的两栖动物，又叫"娃娃鱼"。

Chinese giant salamander (*Andrias davidianus*), also known as "baby fish" in Chinese, is the world's largest and most precious amphibian.

③ 慎出的小麂・贵州铜仁
The timid Reeves's muntjac — Tongren, Guizhou

小麂是麂类中体形最小的一种，活动时非常谨慎，常常做很慢的潜行。中国分布有3种麂，分别是赤麂、黑麂和小麂。

The Reeves's muntjac (*Muntiacus reevesi*), the smallest among all muntjac species, is very cautious by nature and tends to prowl very slowly. There are three species of muntjac in China: Indian muntjac (*Muntiacus muntjak*), hairy-fronted muntjac (*Muntiacus crinifrons*) and Reeves's muntjac.

 晨雾中的白鹤·江西鄱阳湖
White cranes in the morning mist — Poyang Lake, Jiangxi

世界上98%的白鹤都在鄱阳湖越冬，清晨的湖面上浓雾刚刚拉开，兴奋的白鹤不断地鸣叫和舞蹈，犹如一幅雅致的水墨淡彩中国画。

Ninety-eight percent of white cranes (*Grus leucogeranus*) in the world would spend winter in the Poyang Lake. In early mornings when dense mist enshrouding the lake is about to clear away, white cranes will start their excited singing and dancing, presenting the viewers an elegantly-drawn Chinese painting in ink and light color.

凌波仙子·湖北咸宁
A fairy walking lightly on water — Xianning, Hubei

水雉因有细长的脚爪,能轻步行走于睡莲、荷花、菱角、芡实等浮叶植物上,且体态优美,羽色艳丽,又称"凌波仙子"。

Owing to its slender feet, the pheasant-tailed jacana (*Hydrophasianus chirurgus*) can walk lightly on the water lily, lotus, water caltrop, gorgon fruit and other floating plants. With beautiful body posture and colorful feathers, it is also known as the "fairy on water".

多种多样的林鸟·华中区
Diversified forest birds — CCR

本区林鸟种类很多,具有南北过渡特性,多在常绿阔叶与落叶阔叶混交林、常绿阔叶林中活动,也出入于稀树草地、果园、农地、河边与公路边的树上,有时也见于竹林和灌丛。(注:本区的水鸟多为迁徙鸟类,水鸟在其他章节已有论述,本章尽量不再涉及。)图中林鸟依次为:紫啸鸫(图1)、橙腹叶鹎(图2,雌性)、黑头奇鹛(图3)、灰背伯劳(图4)、栗头蜂虎(图5)、蓝喉蜂虎(图6)、黄腿渔鸮(图7)、黑胸太阳鸟(图8,雌性)、眼纹噪鹛(图9)、红尾水鸲(图10)、白腿小隼(图11,白腿小隼是数量稀少的中国最小猛禽,和麻雀差不多大小,能捕食如蝙蝠等小型鸟兽)、斑喉希鹛(图12)、蓝冠噪鹛(图13,蓝冠噪鹛原名黄喉噪鹛,为失踪近一个世纪的世界濒危鸟种,我国仅在江西婺源、云南思茅的狭小地带有发现)。

Many species of forest birds inhabit in the CCR, with transitional characteristics of both those living in the south and north. They are mainly found in deciduous broad-leaved and evergreen broad-leaved mixed forests, evergreen broad-leaved forests, as well as the trees on grass slopes with few trees, orchards, arable lands, river banks and roadsides. And they are occasionally found in bamboo forests and bushes. (Notes: Most of the waterfowls in this region belong to migratory birds, which have been discussed in other chapters and are not covered in this chapter as far as possible.) Forest birds in these photographs are as followed: violet whistling thrush (*Myophonus caeruleus*, Photo 1), orange-bellied leafbird (*Chloropsis hardwickii*, Photo 2, female), dark-headed sibia (*Heterophasia melanoleuca*, Photo 3), grey-backed shrike (*Lanius tephronotus*, Photo 4), chestnut-headed bee-eater (*Merops leschenaulti*, Photo 5), blue-throated bee-eater (*Merops viridis*, Photo 6), tawny fish-owl (*Ketupa flavipes*, Photo 7), black-throated sunbird (*Aethopyga saturata*, Photo 8, female), spotted laughingthrush (*Garrulax ocellatus*, Photo 9), plumbeous water redstart (*Rhyacornis fuliginosa*, Photo 10), pied falconet (*Microhierax melanoleucos*, Photo 11. It is the smallest predatory birds in China, which is about the size of a sparrow, and can prey on small animals such as bats.), chestnut-tailed *Minla* (*Minla strigula*, Photo 12), blue-crowned laughingthrush (*Garrulax courtoisi*, Photo 13. Formerly named yellow-throated laughing thrush, it is one of the most endangered species in the world, which has been missing from the world for nearly a century. Currently, this species is found only in some narrow niches in Wuyuan, Jiangxi and Simao, Yunnan in China.).

◀ 悬停·湖南怀化
Hovering —
Huaihua, Hunan

全世界共有14种太阳鸟，分布于亚洲南部、菲律宾群岛和印度尼西亚。中国有其约一半，分别为蓝喉太阳鸟、绿喉太阳鸟、黑胸太阳鸟、紫颊直嘴太阳鸟、叉尾太阳鸟、黄腰太阳鸟共6种。太阳鸟有细长微弯的嘴和管状的长舌，能和蜂鸟一样空中悬停以吸食花蜜，也被誉为"东方的蜂鸟"。图为蓝喉太阳鸟以悬停的方式吸食花蜜。

There is a total of 14 species of sunbirds in the world, living primarily in areas like the southern Asia, the Philippines and Indonesia. About a half of these species are found to exist in China, including the Gould's sunbird (*Aethopyga gouldiae*), green-tailed sunbird (*Aethopyga nipalensis*), black-throated sunbird (*Aethopyga saturata*), ruby-cheeked sunbird (*Anthreptes singalensis*), fork-tailed sunbird (*Aethopyga christinae*), and crimson sunbird (*Aethopyga siparaja*). With a long, thin, curved bill and a long, tubular tongue, the sunbird can hover in the air for sucking nectar like the hummingbird, for which reason it is also known as the "oriental hummingbird". The Gould's sunbird hovering in this photo is sucking nectar from flowers.

| 生态思考 Ecological Reflection | 长江江豚会成为下一个白鱀豚吗？
Will the Yangtze Finless Porpoise Become the Next Baiji Dolphin?

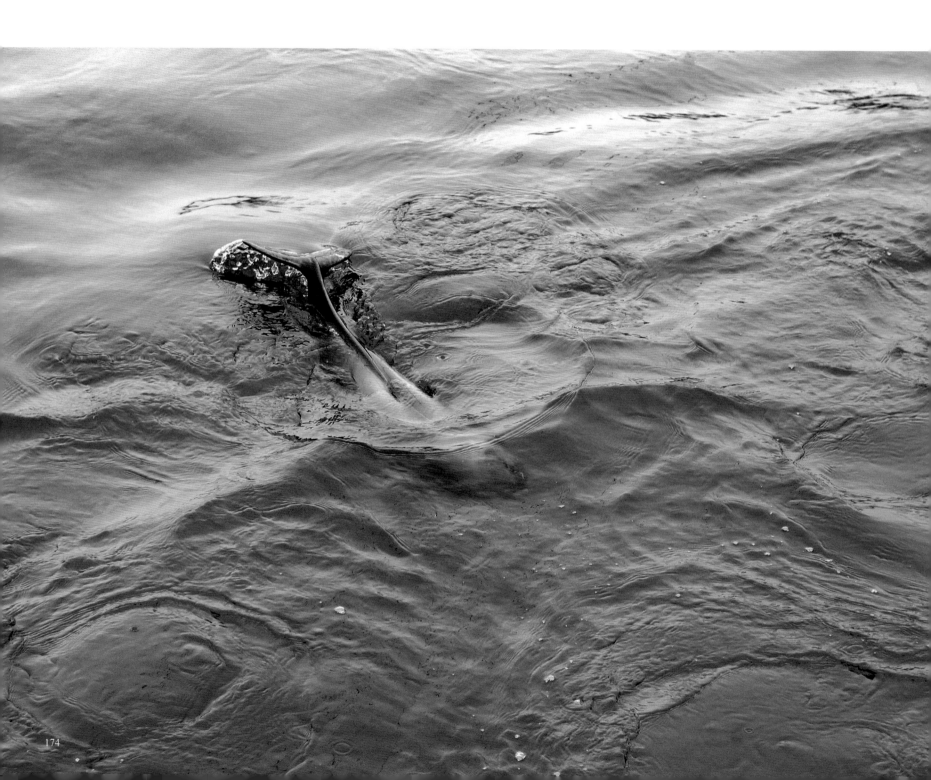

长江江豚已在地球上生存2500万年，被称作长江生态的"活化石"和"水中大熊猫"，现仅分布于长江中下游干流以及洞庭湖、鄱阳湖等区域。长江江豚的主要食物是鱼、虾等。2006年，中国联合6个国家调查长江干流江豚有1200多头，数量已经少于大熊猫，其种群数量还在不断地下降。

造成长江江豚数量减少的原因与造成同是长江豚类的白鱀豚功能性灭绝的原因相同。专家们指出，栖息地恶化、食物资源锐减、长江航运和污染排放是导致流域物种消亡的"四大杀手"。1984年，我作为中国湿地专家组组长协办在岳阳召开的首次"中国湿地保护国际研讨会"之后，专门到中国科学院武汉水生生物研究所去看望中国湿地最濒危的物种——白鱀豚。我看到了悠然游动在水池里"淇淇"的美丽身姿并拍了照片，此事至今久久难以忘怀。2002年7月14日，这头世界上唯一人工饲养的白鱀豚去世，我悲痛不已。没过几年，中、美、英等6国科学家的《2006长江豚类考察报告》称白鱀豚已经功能性灭绝，并警告说，长江江豚正以每年5%~10%的速度在减少，若再不采取有力保护措施，到2020 — 2025年，长江江豚将成为第二个白鱀豚！

专家说：从生物学意义看，作为长江生态系统的关键物种，白鱀豚、长江江豚以及中华鲟、长江鲟、白鲟、鲥鱼等物种几乎同步走向灭绝，这意味着长江生态环境在这一段时期里发生了严重突变。

长江是中国重要的生态宝库，分布有水生生物4000多种，是世界上水生生物多样性最为丰富的河流之一。长江流域有鱼类400多种，其中，特有鱼类近200种。在最新《国家重点保护野生动物名录》（2021年版）中，涉及长江水域的国家一级重点保护野生动物有白鱀豚、长江江豚、白鲟、中华鲟、长江鲟、川陕哲罗鲑、扬子鳄等，国家二级重点保护野生动物有水獭、大鲵、胭脂鱼、金钱鲃、秦岭细鳞鲑等。

现在，白鱀豚、白鲟已经不见踪影，长江江豚危在旦夕，洄游到葛洲坝产卵场的中华鲟亲本目前仅剩20余尾，长江鲟18年来未监测到自然繁殖，但人工繁殖成果可慰——长江鲟子三代（2018年9月）繁殖首获成功，目前，6万余尾长江鲟子三代幼苗生长良好，可持续人工群体已经建立，向自然种群的修复迈进了重要一步……

我曾有一段时间参加主持过几次林业部门的《国家重点野生动物保护名录》的修订，作出了很多艰辛曲折的努力，但由于本人的无能及部门协调机制的僵化，一直没能推动新名录修订出台。长江江豚在2021年新名录公布前还是二十几年前制定的国家二级重点保护野生动物。《国家重点野生动物保护名录》修订拖的时间太长太长，真是无奈、可悲！

2016年1月5日，习近平主席定下基调"要把修复长江生态环境摆在压倒性位置，共抓大保护，不搞大开发"。2018年10月16日，国务院办公厅发出《关于长江水生生物保护工作的意见》强调，要实施以中华鲟、长江鲟、长江江豚为代表的珍稀濒危水生生物抢救性保护行动，内容包括规模化增殖放流，建立驯养繁育基地、接力保种基地、遗传资源库，推动实现野生种群的重建和恢复，还特别强调，要加快提升长江江豚等重点物种的保护等级和自然保护区级别等。2019年年初，农业农村部等三部门联合发文，要求2020年开始实施长江流域重点水域为期十年的常年禁捕。

在我的印象中，以国务院办公厅的名义专门为长江水生野生动物保护发通知实为重大之举，拯救长江生态环境、拯救长江水生野生动物的行动现在已经开始全面铺开。我相信，长江江豚不会成为下一个白鱀豚了。

Known as the "living fossil" and "giant panda in the water" that survives in the ecosystem of the Yangtze River, the Yangtze finless porpoise (*Neophocaena asiaeorientalis asiaeorientalis*) has been living on the Earth for 25 million years. Nowadays, they are found only in the middle and lower reaches of the Yangtze River and the Dongting Lake, Poyang Lake and some other regions. The Yangtze finless porpoise lives mainly on fish and shrimp and so on. According to a survey co-conducted by China and six other countries in 2006, the number of Yangtze finless porpoises that are still surviving in trunk streams of the Yangtze River stands just slightly over 1,200, even less than that of the giant panda. To make it worse, their population is still on a declining trend.

Causes that contribute to the decline in the population of Yangtze finless porpoises are roughly the same with those that led to functional extinction of the Baiji dolphin (*Lipotes vexillifer*), another species of dolphin inhabiting the Yangtze River. Experts

← "江豚时出戏，惊波忽荡漾"·安徽铜陵
The Yangtze finless porpoise over the waves — Tongling, Anhui

这是唐代大文学家韩愈留下的江豚诗，和当时我看到并拍下时的感觉非常相似。
（注：现在已建立的以保护长江江豚为主要保护对象的自然保护区有湖北石首、湖北天鹅洲、湖北洪湖、安徽铜陵、湖南洞庭湖、江西鄱阳湖及江苏南京、江苏镇江等。）

"The Yangtze finless porpoise would often appear above the water to play, whilst the waves were rippled by their activities." This is a line quoted from a poem written by the famous Tang-dynasty poet Han Yu in praising of the Yangtze finless porpoise (*Neophocaena asiaeorientalis asiaeorientalis*). As I spotted the Yangtze Finless Porpoise and captured it down through my camera, I experienced exactly the same feeling as is depicted by this line.
(Notes: Nature Reserves featuring the protection of the Yangtze finless porpoise have been established in Shishou, Tian'ezhou and Honghu in Hubei Province, Tongling in Anhui Province, Dongting Lake in Hunan Province, Poyang Lake in Jiangxi Province, and Nanjing and Zhenjiang in Jiangsu Province.)

长江江豚的微笑·
中国科学院水生生物
研究所
Smiles from the Yangtze finless porpoise — Institute of Hydrobiology, Chinese Academy of Sciences

随着保护力度及科研监测的日益加强，长江江豚应该有一个充满希望的未来。

With ever strengthening efforts in their protection and scientific monitoring, it can be expected that the Yangtze finless porpoise will be guaranteed a promising future.

"Speaking in biologically perspectives, other key species living in the Yangtze River ecosystem, such as the Baiji dolphin, Yangtze finless porpoise, the Chinese sturgeon (*Acipenser sinensis*), Yangtze sturgeon (*Acipenser dabryanus*), Chinese paddlefish (*Psephurus gladius*) and Reeves shad (*Tenualosa reevesii*), are moving towards the verge of extinction at almost the same pace. That means the ecological environment of the Yangtze River has undergone a serious mutation during this period", according to experts.

Serving as an important ecological treasure-house in China, the Yangtze River is teeming with over 4,000 aquatic species, hence ranking it among the rivers that have the richest aquatic biodiversity in the world. There are more than 400 species of fish in the Yangtze River Basin, of which nearly 200 species are endemic. In the newest *List of Key Protected Wild Animals of National Importance* (Version 2021), the Baiji dolphin, the Yangtze finless porpoise, the Chinese paddlefish, the Chinese sturgeon, the Yangtze sturgeon, the Sichuan taimen (*Hucho bleekeri*), the Chinese alligator (*Alligator sinensis*) and other species that live in the Yangtze River are identified for national Class Ⅰ; and the otter (*Lutra lutra*), the giant salamander (*Andrias davidianus*), Chinese sucker (*Myxocyprinus asiaticus*), the golden-line barbel (*Sinocyclocheilus grahami*), *Brachymystax lenok tsinlingensis*, and *etc.* are identified for national Class Ⅱ.

point to "four primary killers" that drive species in this basin onto the verge of extinction, namely: habitat deterioration, food availability that has been worsening alarmingly, shipping along the Yangtze River and mounting pollutants discharged into the waters. In 1984, following the conclusion of the first "International Symposium on Wetland Conservation in China" held in Yueyang, in which I functioned as head of the Chinese expert panel, I specially set aside some time to the Institute of Hydrobiology affiliated to the Chinese Academy of Sciences (CAS) in Wuhan to see the most endangered species in Chinese wetlands — the Baiji dolphin. I saw the beautiful posture of Qi Qi swimming leisurely in the pool and photographed it, which would always be kept in my mind. On July 14th, 2002, this only captive-bred Baiji in the world passed away, which broke my heart. Just several years later, *the Survey Report on Yangtze Dolphins in 2006*, co-written by scientists from China, the U.S. and the UK, reported that the Baiji dolphin was already "functionally extinct" and warned that, calculated on basis of the 5% - 10% annual declining rate, the Yangtze finless porpoise would become next Baiji dolphin by 2020 - 2025 if no effective protection measure is taken to rescue them!

Now, while the Baiji dolpin and Chinese paddlefish have already disappeared completely, the Yangtze finless porpoise is also on the verge of extinct. The number of parent Chinese sturgeons that migrate annually to the spawning grounds of Gezhouba Dam currently stands at just 20. No natural reproduction of the Yangtze sturgeon has been reported over the last 18 years, though promising results have been made in its artificial propagation. Specifically, artificial breeding of third generation sturgeons has been successfully carried out for the first time (in September, 2018), with the population of the third generation young sturgeons standing currently at a self-sustaining level of over 60,000, marking an important milestone on the way

towards the recovery of its wild population.

I have for several times been involved in the revision of the *List of Key Protected Wild Animals of National Importance* under the administration of the forestry sector and contributed my share in this field. Unfortunately, however, owing to my own limited ability as well as rigidity in inter-agency coordination, I have not yet been able to make further breakthrough in the compilation of a completely new list. Over twenty years on, the Yangtze finless porpoise still fell among species for national Class II before the new List (Version 2021) was issued. A thorough revision of the *List of Key Protected Wild Animals of National Importance* has been much too late in coming, which is indeed lamentable!

On January 5th, 2016, President Xi Jinping stressed that top priority must be given to restoring the ecological environment of the Yangtze River, pledging all-out efforts in its conservation by forbidding all large-scale development projects along the river. On October 16th, 2018, the General Office of the State Council issued the *Opinions on the Conservation of Aquatic Life in the Yangtze River*, stressing the need to take action for rescuing and conserving rare and endangered aquatic species represented by the Chinese sturgeon, Yangtze sturgeon and Yangtze finless porpoise. These actions include scaling up breeding and releasing, and constructing domestication and breeding bases, rally breed conservation bases and genetic resource banks so as to promote the reconstruction and rehabilitation of wild populations. Special emphasis was also placed on the needs for upgrading the protection level of such key species as the Yangtze finless porpoise as well as of nature reserves that are established for their sake. In early 2019, the Ministry of Agriculture and Rural Affairs, joined by two other government agencies, issued a document that urged a ten-year all-year-around ban on fishing in key areas in the catchment of the Yangtze River to be put in place in 2020.

In my personal judgment, it is indeed worth noticing that the General Office of the State Council would promulgate a document that was specially aimed at aquatic wildlife protection in the Yangtze River, which marks the formal launch of a full-swing campaign to rescue the ecological environment and aquatic wildlife of the Yangtze River. Under this circumstance, I belive that the Yangtze finless porpoise will not become the next Baiji dolphin.

中华鲟·山东烟台海洋馆
Chinese sturgeon — Yantai Aquarium, Shandong

中华鲟是地球上最古老的脊椎动物之一，是鱼类的共同祖先——古棘鱼的后裔，和恐龙同一时期，是典型的溯河洄游性鱼类，江河大坝的修建对其影响较大，现在已经相当濒危。

The Chinese sturgeon (*Acipenser sinensis*) is one of the oldest vertebrate on the Earth as well as a descendant of Relicanth, the common ancestor of all fishes, whose origin can be traced back to the same era as that of dinosaurs. A typical migration fish, it is now critically endangered due to the great impacts caused by the construction of dams on rivers.

热带雨林·云南西双版纳
Tropical rainforest — Xishuangbanna, Yunnan

树木高大、植被茂密、植物多层、种类繁多的热带雨林是地球上稳定性最高的生态系统,是世界上超过一半的野生动植物物种的栖息地。
Tropical rainforests with tall trees, dense vegetation, multi-layer plants and a wide variety of species are the most stable ecosystem on the Earth and a home to more than half of the world's wild life.

华南区 South China Region

藤蔓世界·云南哀牢山
The vine world — Ailao Mountains, Yunnan
南亚热带湿性常绿阔叶林简直就是各种藤蔓的世界。
The subtropical humid evergreen broad-leaved forest is a world of vines.

南海岛屿·海南西沙
South China Sea Islands — Xisha, Hainan
星罗棋布的南海岛屿上植被多样，潟湖和珊瑚礁里生物多样性丰富。
Scattered all over the South China Sea, the South China Sea Islands are diverse in vegetation, while the lagoons and coral reefs are rich in biodiversity.

红树林中·广西山口
Mangrove forest — Shankou, Guangxi
沿海滩涂及浅海地区的红树林是众多海洋生物栖息的好地方。
The mangroves along coastal shoals and shallow waters are good places for many marine creatures to live.

华南区 为西南区和华中区的以南部分，包括云南、广东和广西三省（自治区）的南部，福建东南沿海一带，以及海南、台湾和南海，涉及云南和广西的南部山地、福建和广东的南部丘陵，以及台湾岛、海南岛和南海诸岛全部，属世界动物地理区划中东洋界的中国南部分，气候区为南亚热带、边缘热带、中热带和赤道热带，是中国最南、最热、降水量最为充沛、光热水条件都非常优越的地区，生物多样性异常丰富。本区动物区系含各类热带—亚热带类型的成分，以东洋型为主，岛屿型和南中国型次之。区域内分布有北白颊长臂猿、海南长臂猿、白头叶猴、蜂猴、坡鹿等数量极少且分布狭窄的特有动物，还有亚洲象、绿孔雀、犀鸟、巨蜥等中国的珍稀动物，南海里有中华白海豚、鲸鲨等海洋动物游弋，绿海龟、玳瑁等在海草床及珊瑚丛中觅食并在海岛沙滩上产卵，各种鹭及鲣鸟等飞翔在海天之间，呈现出的是一派"鸟击长空，鱼翔浅底"美丽生动的热带画面。

该区在生态环境保护方面面临着经济高速发展，工业开发强度日益增强，人口增加，海洋生态系统保护形势紧迫等局面。今后，除了继续强化生态系统和野生动物的保护力度之外，应该将重点放在保护好我国不可多得的热带雨林森林生态系统、海洋生态系统及物种方面。2019年年初，国家批准了《海南热带雨林国家公园体制试点方案》，在新机遇面前，需要大力抓好的薄弱环节是，强化严格的热带雨林保护体系，禁止破坏和侵占现有的各类雨林（除了海南还应包括云南西双版纳、临沧、德宏及西藏东南部河谷区）。建议建立亚洲象国家公园，并下大功夫加强海洋生态系统及珊瑚等海洋生物的保护管理，要创建若干海洋类型国家公园以填补在海洋生态系统上的空缺。在此基础上，大力发展和建设一大批海洋类型的自然保护区和自然公园，要抓好以国家公园为主体的自然保护地体系建设，在"自然保护地分级分类分区"管理上狠下功夫。华南区的自然生态和野生动物保护工作将以此为引领，在中央的大力支持和推动下，经各级地方政府的努力，该区的自然生态系统保护及野生动物保护管理工作将摆脱相对较落后的局面，跃上一个大大的新台阶。

The South China Region (SCR) refers to the region to the south of South-west Region and Central China Region, including the south of Yunnan, Guangdong and Guangxi, the southeast coastal area of Fujian, Hainan, Taiwan and South China Sea, the mountainous areas of Yunnan and Guangxi, the southern hills of Fujian and Guangdong, and Taiwan Island, Hainan Island and all the islands in South China Sea. It belongs to the southern part of the Oriental Realm in world zoogeographic system within China. The climate zones in this region include south subtropic, marginal tropic, mid tropic and equatorial tropic. It is the southernmost, hottest region in China, with the most abundant precipitation, the brightest sunlight, the richest water resources, and great biodiversity. The fauna in this area consist of various tropical-subtropical types, dominated by the Oriental type, followed by the island type and South China type. There are some rare and endemic animals that are only found in narrow part of this land, such as northern white-cheeked gibbon (*Nomascus leucogenys*), Hainan gibbon (*Nomascus hainanus*), white-head langur (*Trachypithecus poliocephalus*), Asian slow loris (*Nycticebus bengalensis*), Hainan eld's deer (*Panolia siamensis*) as well as Asian elephant (*Elephas maximus*), green peacock (*Pavo muticus*), hornbill (*Buceros* spp.), common water monitor (*Varanus salvator*), and *etc*. Marine animals such as Chinese white dolphin (*Sousa chinensis*) and whale shark (*Rhincodon typus*) cruse in the South China Sea. Turtles such as green turtle (*Chelonia mydas*) and hawksbill turtle (*Eretmochelys imbricata*) forage in sea grass bed as well as coral reefs and lay their eggs on the beaches of the islands. All kinds of shearwaters (Procellariidae) and boobies (Sulidae) fly between the sea and the sky, presenting a beautiful and vivid tropical picture of "birds flying across the open sky, fish swimming in shallow water".

In the aspect of ecological environment protection, the area is faced with great pressure caused by accelerating economic and industrial development, increasing population, and there is urgent need to reverse deteriorating trends of its marine ecosystems. In the future, in addition to continuing to strengthen the protection of ecosystems and wildlife, we should focus on protecting the rare tropical rainforest ecosystem, marine ecosystem and species which are precious in China. At the beginning of 2019, the central government approved the *Pilot Program of Hainan Tropical Rainforest National Park*. In the face of new opportunities, the weak link that needs to be vigorously addressed is to strengthen the strict tropical rainforest protection system and prohibit the destruction and occupation of all kinds of existing rainforests (which should also include, in addition to those in Hainan, the rainforests growing in Xishuangbanna, Lincang and Dehong in Yunnan, and the valley areas in the southeastern part of Tibet). It is suggested that an Asian elephant national park should be founded and that great efforts should be made to strengthen the protection and management of marine ecosystem, coral and other marine life, and to create several marine themed national parks to fill the gaps in marine ecosystem conservation. On this basis, we shall vigorously develop and build a large number of marine themed nature reserves and natural parks, do a good job in the development of the a natural protected area system highlighting the central role that national parks play and make great efforts in putting in place a tiered, classified and zoning system for the management of protected areas, which shall guide the work of the conservation of natural ecology and wildlife in the South China Region. With the strong support and promotion of the central government as well as the efforts of local governments at all levels, the conservation of the natural ecosystem and the management of wildlife conservation in the region will shake off a relatively backward situation and leap to a new level.

↑ 孔雀开屏·云南德宏
The peafowl showing its tail — Dehong, Yunnan

雄性蓝孔雀向雌性炫耀自己的美丽，展开了它那五彩缤纷、色泽艳丽的尾屏。孔雀有绿孔雀和蓝孔雀之分，蓝孔雀较常见但是外来物种，为了让公众认识蓝孔雀与受保护的绿孔雀之间的不同，特在此也破例列出。

The male Indian peafowl (*Pavo cristatus*, commonly known as blue peacock) flaunts its beauty to the female, unfolding its colorful and bright tail. There are 2 species of peafowls: green peafowls (*Pavo muticus*) and Indian peafowls. The India peafowl is a more commonly-seen species but exotic to China. In order to help people better understand the difference between the Indian peafowl and the green peafowl that is under protection, the Indian peafowl is also introduced in this book is an exception.

绿孔雀官司的重大社会意义
The Significant Social Implications of Lawsuit Concerning Green Peacocks

孔雀是体型最大的雉科鸟类。在实际生活中，我们会在动物园、"孔雀谷"中看到很多不同颜色的孔雀。其中，蓝孔雀原分布于南亚次大陆，为外来引进物种，而白孔雀则为蓝孔雀的变种。孔雀的种与种之间很容易杂交，我们在动物园等地方看见一些不蓝不白也不绿的孔雀，就是混养杂交的结果。在动物园里我们几乎看不见真正的绿孔雀。

中国原生的孔雀是绿孔雀，仅分布于云南，称"百鸟之王"，在中国的传统文化中，被认为是凤凰的化身。美丽的孔雀被喻为高贵和圣洁之物，自古以来成为文人雅士们讴歌的对象，明清时期三品文官官袍上的"补子"图案是绿孔雀，清代官帽上的孔雀翎也来自绿孔雀。目前，绿孔雀因为种群极度稀少濒危而被列为国家一级重点保护野生动物。

最新调查成果显示，与20世纪90年代初的第一次绿孔雀分布现状调查结果相比，绿孔雀栖息地分布区域已从当年云南的34个县127个镇锐减到今天的22个县33个镇，其种群数量也从当时预估的800~1100只锐减到如今不到500只。绿孔雀濒危的主要原因是栖息地的丧失，如种植的大量经济作物取代了绿孔雀栖息的低山林地和灌丛，还有如非法猎捕行为和村民为了保护庄稼而毒杀绿孔雀，澜沧江、红河等流域的工程开发破坏了绿孔雀求偶觅食的河滩地等。

2017年，一场有关绿孔雀与水电站之间的属地争夺战爆发。其起因是云南楚雄恐龙河自然保护区为了给拟建水电站的大坝淹没区让路等原因，于2008年、2010年先后3次进行过面积调整，而其中戛洒江水电站的大坝蓄水淹没区带来了孔雀栖息地之一的河滩会随之消失的危害，还会淹没国家一级重点保护野生植物陈氏苏铁以及千果榄仁、多种兰科等濒危植物。于是，中国的三家民间环保组织联名给环境保护部（现生态环境部）发去了紧急建议函，之后，又将负责具体开工建设戛洒江水电站的水电公司和环评单位告上法庭，一座投资数十亿元的水电站因保护绿孔雀和陈氏苏铁而被叫停。一份程序合法但内容和实际情况有差异的环评报告让绿孔雀和水电站之间的这场官司陷入困境，此事也成为了当今社会关注的热点之一。2020年3月20日，昆明市中级人民法院判决停止该项目的建设，要求不得截流蓄水，不得对淹没区内植物进行砍伐，后续处理待后。

无论将来最终结果如何，我更看重的是事情本身所反映出来的重大社会意义。这是2014年修订的《中华人民共和国环境保护法》实施后，首例由社会组织提起的野生动物保护预防性环境民事公益诉讼。它的意义还在于：是因为保护野生动物，是因为要提前预判生态损失，是因为公益诉讼不再仅仅是由检察机关而是由社会民间组织来提起。这是我国依法治国、依法保护野生动物进程中具有里程碑意义的案件，事情本身就值得大大点赞！

The peacock is the largest bird of the pheasant family. In real life, we will see many peacocks of different colors in zoos and "peacock valleys". Among them, the blue peacock, an exotic species from which its white peacock variant is derived, was originally distributed in the South Asian subcontinent. Peacocks are easy to hybridize. The peacocks we often see in zoos and other places that are neither pure blue, pure white nor pure green color are the result of hybridization. We can rarely see real green peacocks in the zoo.

The native peacock in China is the green peacock that is found only in Yunnan, known as the "king of birds". It is regarded in traditional Chinese culture as the embodiment of phoenix. The incarnation of nobility and holiness hence often celebrated and worshipped in classical literature work and various folklores since ancient times. The pattern on the "bu zi" attached to the official robes of the third-tier civil officials during the Ming and Qing dynasties is the peacock, and the feather decorated on typical official hats during the Qing Dynasty were also taken from this bird. At present, the green peacock is listed as Class I Key Protected Wild Animal of National Significance due to its extremely small population and highly endangered state of existence.

According to the results obtained from the latest survey, compared with the early 1990s when the first survey on the distribution of green peacocks was carried out, the habitat of green peacocks have sharply shrunk from an area that covered 127 towns in 34 counties in Yunnan Province back then to an area that covers merely 33 towns in 22 counties today, and their population has also sharply decreased from then estimated 800 – 1,100 to less than 500 today. The main reasons for the endangered green peacock are the loss of habitats, such as the cultivation of mass economic crops to replace the low mountain woodland and shrub inhabited by green peacocks, illegal hunting and killing of green peacocks by villagers in order to protect crops. Besides, engineering development along the Lancang River, the Honghe River and other basins has also inflicted tremendous harms to the waterside niches where green peacocks court and forage.

In 2017, a territorial dispute related with the protection of green peacocks and the needs for building hydropower stations broke out. The reason is that, in order to accommodate to the construction of dams used by a planned hydropower station, the Dinasour River Nature Reserve of Chuxiong in Yunnan Province had readjusted its boundaries for three consecutive times in 2008 and 2010. The planned hydropower station on the Jiasajiang River would not only inundate areas that the peacocks inhabits, but also pose danger to the survival of *Cycas chenii*, a precious wild plant listed for Class I Key Protected Wild Plants of National Significance and some endangered plants including *Terminalia myriocarpa* and some orchids. Therefore, three non-governmental environmental protection organizations in China jointly filed an emergency proposal to the Ministry of Environmental Protection (present Ministry of Ecology and Environment), in addition to suing the hydropower company as well as the environmental assessment agency responsible for the construction of the Jiasajiang Hydropower Station. A multibillion-yuan hydropower station was halted for protecting the green peacock and *Cycas chenii*. An environmental impact assessment report, which has been produced through a legitimate procedures but whose content deviates notably from the real situation on the site, puts the lawsuit concerning the protection of green peacocks and the construction of hydropower stations in a dilemma, and makes up headline in the media. On March 20, 2020, the Kunming Municipal Intermediate People's Court judged to stop the construction of hydropower station, ban damming and storing water, ban felling the plants within the submerged area and pend the follow-up disposal.

Whatever the final result of the lawsuit is, what I value more is the social implications of the event itself. This is the first civil lawsuit filed by Chinese environmental NGOs that is aimed to prevent potential damages to the protection of wildlife in the wake of the implementation of the revised *Environmental Protection Law of the People's Republic of China* in 2014. Its significance also lies in the following aspect: it is initiated by NGOs rather than by procuratorial organs of the country for the sake of wildlife protect and raised on the basis of anticipated damages to the eco-environment. This event marks an important milestone on China's journey towards ruling the country by law and protecting wildlife by law, thus shall be highly commended.

↑ 白孔雀·广东长隆
White peacock — Chimelong, Guangdong

现在大量饲养或展出的孔雀都是蓝孔雀或其变种（包括白孔雀）。

Peacocks that are bred and widely displayed currently are all Indian peacocks or their genetic variants (including white peacocks).

← 高贵的绿孔雀·云南玉溪
Noble green peacock — Yuxi, Yunnan

绿孔雀为中国特有雉类，现仅生活在云南南部，异常珍贵。在中国的传统文化中，绿孔雀被认为是"百鸟之王"。

The green peacock (*Pavo muticus*) is a unique pheasant in China. It only lives in the southern Yunnan, extremely rare and precious. In Chinese traditional culture, the green peacock is considered the " king of all birds".

热带雨林的犀鸟故事
The Story of Hornbills in Tropical Rainforests

飞翔的冠斑犀鸟·云南盈江
Flying Malabar pied crowned hornbill — Yingjiang County, Yunnan

犀鸟由于翅展很大，飞行起来扇动翅膀的声音能够回荡在森林上空。听到声音就能够判断其飞行的大概方向。

The wingspan of hornbill is so large that the sound of fluttering wings can reverberate over forests. One can judge the general direction which the bird is flying by listening to the sound of its wing fluttering.

犀鸟属典型的热带森林鸟，中国的犀鸟主要生活在云南西部、南部以及广西南部，共有5个种（均为国家一级重点保护野生动物）：双角犀鸟、冠斑犀鸟、白喉犀鸟、棕颈犀鸟和花冠皱盔犀鸟。云南盈江是中国唯一记录有5种犀鸟的地方，也是犀鸟分布的最北缘。在那么高的纬度居然还有热带雨林和犀鸟的原因，是南北走向的高黎贡山阻挡了印度洋来的暖湿气流，在西坡营造了特殊的局部沟谷热带雨林气候。中国最大的一片由龙脑香科植物组成的热带雨林就分布在这里。

犀鸟主要栖息于海拔1500米以下的低山和山脚平原的常绿阔叶林中。春末夏初，犀鸟经长时间调情并交配后，选择在高高的大树洞里产卵，当雌鸟卧进树洞里孵卵后，雄鸟就衔泥将洞口封闭，只留下一条窄长的孔，以避免蛇、猴子等天敌对雌鸟的伤害。在雌鸟孵卵直到雏鸟羽毛长出的整个过程中，全由雄鸟从条孔中给雌鸟喂食。雄鸟每天远寻近觅，劳碌奔波于森林与"家庭"之间，白天忙过，夜晚还要站岗放哨，非常地辛苦！两个多月后，小犀鸟破壳而出，由雄鸟再喂养数天后，雌鸟就用嘴把洞口啄开，为自己解除"禁闭"，并开始和雄鸟一起哺育雏鸟。有那么一个美丽而凄凉的传说：一对犀鸟中，如有一只死去，另一只也绝不苟且偷生或另寻新欢，而是在忧伤中绝食而亡，因此，也将犀鸟称为"钟情鸟"。

2019年4月，我亲历了一次"犀鸟的世界难题"故事，揭开了这个美丽而凄凉传说的神秘面纱。故事是从一个月前的3月5日开始的，雄性双角犀鸟依依不舍地把雌鸟送进了那个多年的树洞爱巢，将洞口封好后，开始每天不辞辛苦地为雌鸟送去食物。可是在进洞才一月余的4月7日，雄鸟却再也没有如约而至，4月9日，村民在路边发现它已经死亡。雄鸟已经死了3天，树洞里面等待喂食的雌鸟怎么办？这个突发事件牵动了无数自然生态保护者的心，除了应及时确定雄鸟的死因外，洞里的雌鸟如何救护就成了我们必须要尽快解决的难题。届时的我们，头脑里只有那个凄美的传说而没有任何经验和知识，国内也没有犀鸟方面的专家可以咨询。怎么办？我当时及时采取措施，通过香港嘉道理鸟类专家的渠道，联系了IUCN犀鸟工作组及泰国的犀鸟专家——"犀鸟妈妈"，请他们提出救护建议。但是一天过去了，回复是没有这方面的成熟建议，这还真就成了一道世界性的难题了。在等待的现场，我们设想了各种救援方案，准备了攀爬大树的应备工具及无人机、绳梯等（树洞离地面近30米高），安排了善于爬树的当地民族小伙，打算用人工喂养的方式来救护洞里的雌鸟。正在大家万分焦虑紧张地准备时，雌鸟竟然自己破洞而出了！

随后的跟踪发现，这只雌性双角犀鸟很快恢复了体力，又自由地翱翔在了热带雨林的上空，并在原树洞中"改嫁育子"！这只雌犀鸟用活生生的事实改写了神秘的传说，解决了一道世界难题，更用行动诠释了大自然的生存法则，给我们上了一堂生动的"适者生存"的科学课。

The hornbill is a typical tropical forest bird. In China, they mainly live in the western and southern Yunnan and the southwestern Guangxi. There are five species in total that are all listed as Class I Key Protected Wild Animal of National Significance: great hornbill (*Buceros bicornis*), Malabar pied crowned hornbill (*Anthracoceros albirostris*), Austen's brown hornbill (*Anorrhinus austeni*), rufous-necked hornbill (*Aceros nipalensis*) and wreathed hornbill (*Rhyticeros undulatus*). Yingjiang in Yunnan Province is the only place in China whereas all the five species of hornbills have been recorded, and it is also the northernmost edge of hornbills' distribution. The reason why there are rainforests and hornbills at such a high latitude is that the north-south Gaoligong Mountain blocks the warm and humid air flow from the Indian Ocean, and therefore creates a special local valley tropical rainforest climate on the west slope of the mountain where the largest tropical rainforest composed of Dipterocarpaceae plants in China is located.

Hornbills mainly inhabit the evergreen broad-leaved forests in the low mountain and foothill plains below 1,500 meters in altitude. At the end of spring and the beginning of summer, after a long period of courtship and mating, hornbills would start to lay eggs in high big tree holes. When the female stays in tree holes for incubation, the male would block the entrance to the holes with mud, leaving just a long narrow crack through which food can be supplied, to prevent predators like snakes and monkeys from getting into the hole and hurting the female. Throughout the entire period that ends only when their nestlings are fully fledged, the male bird would be responsible for feeding the female bird and her nestlings through the narrow crack. In order to search for food, the male bird forages between the forests and their nest daily, busy in the daytime and standing on guard at nights. Two months later, hornbill nestlings crack their shells to come out, the male would go on feeding the female and her nestlings for a few days, before the female removes the block at the entrance to the nest with her beak to free herself from confinement and joins her husband in raising their offspring. It is alleged that when one of the couple dies, the other one will never go on with its lonely living or find a new spouse. Instead, the heart-broken one left behind would choose to starve itself to death. For this reason, hornbills are often hailed as "love birds".

In April 2019, I personally experienced a "world puzzle of hornbills" that unveil the myth concerning the heart-rending love stories of these birds. The story began about one month ago on March 5, when the male hornbill reluctantly sent her spouse in their age-old nest of love in the tree and started his painstaking daily routine for feeding and guarding the safety of the female bird. Everything went on smoothly for around one month until April 7, when the male bird failed to arrive at their nest as scheduled and was found dead by the roadside two days later on April 9. It has been three days since the male died, what would happen to the female bird that was still waiting for food in their nest? This emergent situation raised the concern of countless conservationists who, in addition to investigating on the cause of the male bird's sudden death, had to come up with immediate measures to rescue the starving female in the nest. At that time, we had, except for their alleged story, neither previous experience to draw on nor hornbill experts to consult from domestically. What should we do? Through the help of bird experts from Kadoorie Farm & Botanic Garden based in Hong Kong, I contacted the IUCN Hornbill Working Group and the Thai hornbill expert — "Mother Hornbill" for some useful tips on how to address the urgent problem. But the answer arrived one day later was that they did not have a ready solution either. It had indeed become a great puzzle for the entire world! Various contingent plans were proposed on the site, with all necessary tools like drones, rope ladders, *etc.* (the hole is nearly 30 meters high from the ground) prepared so that someone skillful in climbing trees could climb up to feed the female birds artificially. While all these hustling were going on, to our surprise, the female bird in the nest removed the block on her own efforts and flew out!

We tracked this great hornbill closely afterwards and found that she soon recovered her strength, were soaring freely over the tropical rainforests, as well as remarried and bored offsprings in the former tree hole! This female hornbill not only solved on her own the big problem that puzzled the whole world and cracked the myth concerning their love story, but also, through her brave action in dismantling the barrier, gave us a telling lesson about life in the mysterious natural world — "survival of the fittest".

调情·云南盈江铜壁关
Courting — Tongbi Pass, Yingjiang County, Yunnan

花冠皱盔犀鸟在我国仅见于云南盈江县一带，羽冠栗色、有一黄色喉囊的为雄鸟。繁殖期即将到来，两夫妻相互调情，一改平日总是雄性向雌性献殷勤的惯例，雌性也用果子回报雄性。

In China, the wreathed hornbill (*Rhyticeros undulatus*) is only found in Yingjiang County, Yunnan Province. The male has a chestnut crown and a yellow throat pouch. As the breeding period is approaching, this couple is courting to each other. Whereas normally it is for the male to please the female, a change can be observed during the courtship, with the female occasionally seen to return the favor by treating her male partner with fruits and nuts.

喂蛇·云南盈江石梯
Feeding on a snake — Stone Ladder, Yingjiang County, Yunnan

雌性双角犀鸟躲在树洞中孵育小鸟已经2月有余，身体需要更多的能量补充，雄性犀鸟除了来回奔波给雌鸟送来浆果以外，还不时添加点蜥蜴、蛙类等肉食，这次捕ого1米左右长的蛇（过树蛇）准备喂入，十分罕见。这只犀鸟就是前文"热带雨林的犀鸟故事"里意外死亡的男主人翁，此图是2017年4月拍摄到的。

It has been more than two months since the female great hornbill (*Buceros bicornis*) stayed in the tree hole to incubate young chicks, and their body needs more energy to supplement. In addition to rushing back and forth to bring berries to the female, the male also adds some meat such as lizards and frogs occasionally. This photo taken in April, 2017 is very rare for that what he has brought home for her this time is a 1-meter-long snake (painted bronzeback), which is indeed very unusual. This is just the male bird we previously mentioned in *The Story of Hornbill in Tropical Rainforests* that died by accident while his partner was in incubation.

两岸交流的"明星"
Stars of Cross-strait Exchanges

挑战·山东威海刘公岛
Challenge — Liugong Island of Weihai, Shandong

对于生于热带的台湾梅花鹿而言,在威海过冬是一项挑战。
(注:台湾鬣羚和台湾梅花鹿均属于台湾的原生动物,因此本书将其放在华南区。)
For the tropical-born Taiwan sika deer (*Cervus nippon taiouanus*), wintering in Weihai is a tough challenge.
(Notes: Both Taiwan serow and Taiwan sika deer are native species of Taiwan, China. Therefore they are introduced in SCR.)

2008年11月6日,海峡两岸关系协会会长陈云林和海峡交流基金会董事长江丙坤在台北举行了两岸互赠大熊猫、梅花鹿和长鬃山羊(现称"台湾鬣羚")仪式。而后,大陆赠台大熊猫"团团""圆圆"于2008年12月23日乘专机从成都前往台北,落户在木栅动物园。为迎接大陆赠台大熊猫的这份特殊的礼物,动物园特地为大熊猫"团团"和"圆圆"设立了网站、开了博客。大熊猫深受台湾民众喜爱,掀起了一场"大熊猫热"。至此,大陆同胞向台湾同胞赠送大熊猫的心愿与台湾同胞们期待在台湾目睹大熊猫风采的心愿,在这一天终于圆满实现了。当时的我也抑制不住欣喜的心情,在《人与自然》杂志上接连发表了《熊猫"内幕"》《熊猫:光荣和梦想》和《卧龙盛事》三篇文章,讲述了大熊猫赠台前前后后的故事。2013年7月6日晚,大熊猫"圆圆"产下经人工受孕后得到的宝贝,取名"圆仔"。 2019年7月6日,台北动物园为"圆仔"举办了6岁"生日宴",大熊猫"圆仔"享用生日蛋糕的图片在网上再一次引起轰动。

台湾鬣羚是台湾特有动物,也是台湾唯一的野生牛科动物,拥有较高的跳跃能力,在台湾所有的哺乳类动物中,堪称"轻功第一高手"。台湾赠送大陆的台湾鬣羚取名为"喜羊羊"和"乐羊羊",并于2012年年初在山东威海刘公岛生下雄性一仔,取名"甜甜"。台湾梅花鹿是台湾的特有亚种,野生族群来源为台湾垦丁国家公园,垦丁是梅花鹿种原始分布的最南端。台湾赠送大陆的梅花鹿被命名为"繁星"和"点点",来时怀有一胎,当年冬天出生,取名为"晶晶"。现在,这个家庭已经繁衍到十多头了。

如今,在台湾出生的大熊猫"圆仔"和在大陆出生的台湾鬣羚"甜甜"都已经到了性成熟阶段。希望两岸各界一起努力,在和平友好的气氛中共搭鹊桥,为"圆仔"和"甜甜"早择配偶,以了两岸人民的一致心愿!

On November 6th, 2008, Chen Yunlin, President of Association for Relations across the Taiwan Straits and Jiang Bingkun, Chairman of the Strait Exchange Foundation, held a ceremony in Taipei to exchange giant pandas, sika deer and Taiwan serows. On December 23rd, 2008, giant pandas, Tuan Tuan and Yuan Yuan, were flown from Chengdu to Taipei by the special plane and settled in Muzha Zoo. In response to the special gift from the mainland, the zoo has opened a website and a blog that were specially dedicated to Tuan Tuan and Yuan Yuan. They were so popular in Taiwan that a "panda fever" has set off on the Island. The wish of the mainland compatriots to present the panda to the Taiwan compatriots and the wish of the Taiwan compatriots to see the living pandas in Taiwan have finally come true. At that time, I couldn't help my happiness and published three articles successively in the *People and Nature*: *Inside Story of the Pandas*, *Panda: Glory and Dream*, and *A Historic Event in Wolong*, through which a detailed narration concerning the trip of the pandas to Taiwan is presented to readers. On the evening of July 6, 2013, Yuan Yuan gave birth through artificial pregnation to a cub that was named Yuan Zai. On July 6th, 2019, Taipei Zoo held a birthday party for the 6-year-old Yuan Zai. The picture that depicts Yuan Zai enjoying his birthday cake went viral again on the Internet.

The Taiwan serow (*Capricornis swinhoei*) is an animal endemic to Taiwan, and also the only wild bovine in Taiwan.

"甜甜"成年了·
山东威海刘公岛
The maturity of "Tian Tian" — Liugong Island, Weihai, Shandong

来自宝岛台湾的台湾鬣羚"乐羊羊"和"喜羊羊"已经适应了北方寒冷的气候，快乐地生活着并繁育了后代"甜甜"，如今"甜甜"已经性成熟了。

The Taiwan serow (*Capricornis swinhoei*) "Le Yangyang" and "Xi Yangyang" from Taiwan have now adapted to the cold climate in the north, and live happily and bred their offspring "Tian Tian" who has now reached sexual maturity.

Highly adept in jumping, it is touted as "the greatest master of Qing Gong (a skill of Chinese Martial arts good at jumping high)" among all the mammals in Taiwan. The Taiwan serow presented to the Mainland by Taiwan were named Xi Yangyang and Le Yangyang. In early 2012, they gave birth to a male cub named Tian Tian on the Liugong Island, which is situated in Weihai, Shandong Province. The Taiwan sika deer is an endemic subspecies in Taiwan. They are originated from the wild population in Kenting National Park, the southernmost place where wild population of the sika deer are found. A couple of sika deer that Taiwan presented to the Mainland were respectively named Fan Xing and Dian Dian who was already pregnant before arriving at the Mainland and gave birth in the winter of the same year to a calf named Jing Jing. As of the present, the family has grown to a large group with over 10 deer.

Now, the Taiwan-born giant panda, Yuan Zai and the Mainland-born Taiwan serow, Tian Tian, have both reached sexual maturity. It is hoped that all concerned parties on both sides of the Taiwan Strait will work closely in their concerted efforts to choose for Yuan Zai and Tian Tian their respective spouse so as to fulfill the common aspiration of people across the Strait!

海洋生物保护行动
Marine Life Conservation Action

游向大海·海南西沙
Swimming to the sea — Xisha, Hainan

小绿海龟终于走完了从出生的沙窝到海边的艰难历程，一阵海浪过来，小海龟被海水托起，随浪游进大海的深处。

The little green turtle (*Chelonia mydas*) is finally through with the hard journey that extends from its birth place in the sand to the seaside. With a wave coming over, the little green turtle is lifted up and swim into the deep sea with the wave.

欢腾的南海·海南南沙
Jubilant South China Sea — Nansha, Hainan

热带点斑原海豚成群结队地跳出海面，不断地追逐着航行的船。

The pantropical spotted dolphins (*Stenella attenuata*) are jumping out of the sea in droves, constantly chasing the sailing ships.

中国的南海中生长着很多著名的海洋生物。

中华白海豚是近岸海洋生态系统的旗舰物种和指示物种，位于近岸海域食物链的顶端，具有重要的生态、科研和文化价值。中华白海豚在福建和广东一带被渔民尊称为"妈祖鱼"。

海龟作为和恐龙同一时代并存活至今的物种，在生物进化史上有着不可替代的位置，被称为海洋"活化石"。在中国的传统文化中，海龟是吉祥、长寿的象征。中国有世界上现存海龟属8种中的5种，分别为：绿海龟、玳瑁、蠵龟、丽龟和棱皮龟。

近年来，由于海洋海岸工程的数量和强度日益增加，滨海湿地退化及海岸滩涂的大量减少，大量的陆源排污、过度捕捞等原因，海洋生物资源锐减，生态系统功能退化，造成了以中华白海豚及海龟等为代表的海洋生物栖息地的萎缩和严重破碎化。为了保护重要海洋物种及其栖息地，我国先后设立了若干个海洋生物自然保护区。近年来，农业农村部加强了海洋生物的保护力度，先后发布《中华白海豚保护行动计划》和《海龟保护行动计划》，要求建立和完善监测、评估、预警、救护和公众宣传工作体系，提出的中期目标是使物种衰退趋势得到有效遏制，终期目标是使物种种群数量保持稳定或小幅回升，栖息地破碎化现象得到有效缓解。为了更广泛地动员科研、教育、保护区等基层组织都投入到保护海洋生物的行动中，还分别成立了中华白海豚保护联盟和中国海龟保护联盟。

我国有300万平方千米的管辖海域，有6500多个岛屿和数以千计的大小礁盘，有包括鲸鲨、海龟、海豚等在内的数十种海洋大型生物，确实需要花大力气来改变海洋生物目前的生存状况。目前，在我国海洋生态系统中还没有国家公园，衷心希望以海洋国家公园为主体的海洋自然保护地体系能够尽快建立，为保护好我国海洋生态系统及海洋野生动物提供有力的保障。

The South China Sea is home to many famous marine creatures.

For example, the Chinese white dolphin (*Sousa chinensis*) which sits at the top of the food chain in offshore waters, and has important ecological, scientific and cultural values, is the flagship species and indicator species of coastal marine ecosystems. Chinese white dolphins are honored as "Matsu fish" by fishermen in Fujian and Guangdong provinces.

As a species that lived in the same era as dinosaurs and still alive today, sea turtles have an irreplaceable position in the history of biological evolution, and are called "living fossils" of the ocean. In traditional Chinese culture, turtle is regarded as a symbol of auspiciousness and longevity. Among the eight sea turtle species existing in the world, five are found in China, namely: green turtles (*Chelonia mydas*), hawksbill turtles (*Eretmochelys imbricata*), loggerhead sea turtles (*Caretta caretta*), olive ridley (*Lepidochelys olivacea*) and leatherback sea turtles (*Dermochelys coriacea*).

In recent years, due to increasingly amount and intensity of coastal engineering projects, degradation of coastal wetlands and reduction of coastal tidal flats, coupled with increased outpours of land-sourced wastes, exploitative fishing practices, and many other reasons, the marine biological resources have declined sharply and ecosystem function have degraded alarmingly. As a consequence, habitats for marine lives represented by Chinese white dolphins and sea turtles have greatly shrunk and fragmented. In order to protect the important marine species and their habitats, China has successively established a number of nature reserves for marine life conservation. In recent years, the Ministry of Agriculture and Rural Affairs has strengthened protective efforts for marine lives, and issued the *China White Dolphin Conservation Action Plan* and the *Sea Turtle Conservation Action Plan* consecutively, urging all parties concerned to put in place a mechanism that integrates the monitoring, assessment, early warning, rescue and public education activities into a well streamlined system. According to the plans, the mid-term goal is to effectively curb the trend of the population decreasing, while the final goal is to keep the population of protected species stable or moderately increased, with the problem of habitats fragmentation notably mitigated. The China White Dolphin Conservation Union (CWDCU) and the China Turtle Conservation Union(CTCU) have also been set up respectively in a bid to mobilize such grassroots organizations as scientific research institutions, education organizations, and nature reserves to participate in the protection of marine life.

China has three million square kilometers of jurisdictional waters, more than 6,500 islands and thousands of reefs. There are

dozens of large marine creatures, including whale sharks, sea turtles, and dolphins. Great efforts are needed to change the current living conditions of marine lives. At present, no national park that features the marine ecosystems is yet available in China. We sincerely hope that a marine protected area system that centers on marine national parks can be established as soon as possible, so as to provide robust guarantee for the protection of the marine ecosystems and marine wildlife in the country.

① 玳瑁・海南西沙
Hawksbill turtle — Xisha, Hainan

玳瑁是杂食性动物，主要生活在浅水礁湖和珊瑚礁区，和海龟一样在海岸沙滩挖穴产卵。

The hawksbill turtle (*Eretmochelys imbricata*) is an omnivorous animal that lives mainly in shallow water lagoons and coral reefs. Like other sea turtles, the hawksbill digs holes and lay eggs on the beaches of the coast.

③ 翻车鱼・海南西沙
Ocean sunfish — Xisha, Hainan

翻车鱼为大型大洋性鱼类，主要以水母为食，因喜欢侧身游泳和将肚皮翻起来晒太阳而得名。

The sunfish (*Mola mola*) is a large oceanic fish, which mainly feeds on jellyfish. It is named as such because it has the habit of swimming sideways and turning its belly up to bask in the sun.

② 中国鲎・广东雷州
Horseshoe crab — Leizhou, Guangdong

中国鲎生活在浅海沙质海底，也到海滩上活动，其血液为蓝色，雌体比雄体大，是世界上最古老的动物之一，逃过了5次地球生物大灭绝，有"生物活化石"之称。图为红树林里的一对中华鲎。

The blue-blooded horseshoe crab (*Tachypleus tridentatus*), living on sandy sea floor and moving about on beach, is one of the oldest animals in the world. Females are larger than males. Having survived five massive extinctions that struck the Earth, it is often hailed as "living fossil". The photo dispays a couple of horseshoe crab in mangrove forest.

④ 绿海龟・海南三沙北岛
Green turtle — Sansha North Island, Hainan

绿海龟分布于热带及亚热带海域，除了上岸产卵外，终其一生都在大洋中度过。为了保护已经在地球上生活了约2亿年的海龟，每年5月23日被定为"世界海龟日"。

Green turtles (*Chelonia mydas*) are distributed in tropical and subtropical waters, and they spend their whole life in the ocean except for laying eggs on the shore. In order to protect the turtles that have lived on the Earth for about 200 million years, May 23rd is designated as "World Turtle Day".

珊瑚和鱼群・海南三亚
Corals and fish — Sanya, Hainan

珊瑚虫在生长过程中能吸收海水中的钙和二氧化碳，分泌出石灰石（也称珊瑚石）。珊瑚生态系统是海洋生物多样性最为丰富的地方，是各种鱼虾及底栖生物的天堂。

Corals can absorb calcium and carbon dioxide in the sea water and secrete limestone (also known as coral stone) as they grow. The coral ecosystem is the place with the richest marine biodiversity, and it is a paradise for various fish, shrimps and benthic organisms.

出水神灵·广东伶仃洋
The deity out of the water — Lingdingyang, Guangdong

中华白海豚身体修长，全身呈象牙色或乳白色，背部散布有细小的灰黑色斑点是其特征。

The Chinese white dolphin is slender. Its body is ivory or milky white, with its back dotted with small gray black spots.

逐浪捕鱼·广西三娘湾
Fishing in waves — Sanniang Bay, Guangxi

图为中华白海豚在逐浪捕鱼，搅动的鱼群也激发了海鸥强烈的捕食欲望，紧随不舍。中华白海豚有"海上大熊猫"之称，其主要栖息地为红树林水道、海湾等浅海地区。

This picture shows that, in the waves the Chinese white dolphins (*Sousa chinensis*) are fishing in the waves. The agitated fish group stimulates the seagulls' strong desire for food, and they are reluctant to give up. Known as "giant pandas of the sea", the Chinese white dolphins mainly live in shallow waters such as mangrove waterways and bays.

世界最大的鱼·广东长隆海洋世界
The world's largest fish — Chimelong Ocean World, Guangdong

鲸鲨身体庞大，全长可达20米，靠在海水中滤食大量浮游生物和小型鱼类为生，中国各海区夏、秋季节都有其分布。人类捕捞是其数量减少的重要原因。

The whale shark (*Rhincodon typus*) is huge, with a total length of up to 20 meters. It lives by filtering and eating a large number of plankton and small fish in the sea water. It is distributed in various sea areas of China in summer and autumn. Fishing by human is the key causes for its declining population.

求爱・云南盈江
Courtship — Yingjiang County, Yunnan

灰孔雀雉雄鸟在向雌鸟求爱时，会尽量展开双翅，尾巴上扬，以炫耀自己的美丽和威武。孔雀雉属在中国有灰孔雀雉和海南孔雀雉2种，都分布在华南区，是分布区域狭窄、数量稀少、非常珍贵的鸟类。

When courting the female, the male grey peacock pheasant (*Polyplectron bicalcaratum*) will try to spread its wings and raise its tail to show off its beauty and power. There are two species of peacock pheasants in China, namely the grey peacock pheasant and the Hainan peacock pheasant, both of which are extremely precious birds in SCR that are found only in narrow niches and are pretty small in populations.

多种多样的鸟类・华南区
A variety of birds — SCR

图中依次为：鹤中珍品——赤颈鹤・广东长隆（图1）、3个脚趾的翠鸟——三趾翠鸟・云南盈江（图2）、山鹧鸪中的明星——红喉山鹧鸪・云南保山（图3）、空中捕鱼（飞鱼）绝技——红脚鲣鸟・海南西沙（图4）、浪中渔——岩鹭・香港大屿山（图5）、好斗鸡——棕胸竹鸡・广东肇庆（图6）、家鸡祖先——红原鸡・云南铜壁关（图7）

The birds in the photos in turn are: the rarities among cranes — Sarus crane (*Grus antigone*), Chimelong of Guangdong (Photo 1); the kingfisher with three toes — oriental dwarf-kingfisher (*Ceyx erithaca*), Yingjiang of Yunnan (Photo 2); the star of mountain partridgerufous — throated partridge (*Arborophila rufogularis*), Baoshan of Yunnan (Photo 3); fishing from the air-red footed booby (*Sula sula*), Xisha of Hainan (Photo 4); fishing in waves — Pacific reef heron (*Egretta sacra*), Lantau Island of Hong Kong (Photo 5); a belligerent cock — mountain bamboo-partridge (*Bambusicola fytchii*), Zhaoqing of Guangdong (Photo 6); the ancestor of domestic chicken — red jungle fowl (*Gallus gallus*), Tongbi Pass of Yunnan (Photo 7).

夜行者·云南盈江石梯
Night fliers — Stone Ladder, Yingjiang County, Yunnan

黑顶蛙口夜鹰在中国仅见于云南德宏，非常稀少，白天多躲藏在密林中，晚上才开始活动和捕食，因体色与树枝较相似而很难被发现。

The Hodgson's frogmouth (*Batrachostomus hodgsoni*) is only found in Dehong, Yunnan Province. It is a very rare species which normally hides in dense forest during the day and starts to hunt at night. It is difficult for people to come across this bird because its body color is similar to branches.

❶ 新的希望·云南盈江
New hope — Yingjiang County, Yunnan

河燕鸥在中国只在云南盈江境内有记录，目前仅有7只成鸟。2019年春，在盈江的江心滩上有3个繁殖巢中正在发生孵育行为。图为亲鸟从江里抓鱼来喂躲在翅膀下面的幼鸟。

The river tern (*Sterna aurantia*) is only found in Yingjiang County, Yunnan Province, and only seven adult birds are recorded at present. In the spring of 2019, three nests in which female birds incubate were found on the bank of the Yingjiang River. The photo shows the birds catching fish from the river to feed the young chicks hiding under the mother's wings.

❸ 飞鸟出洞·云南盈江石梯
Taking off — Stone Ladder, Yingjiang County, Yunnan

红腿小隼和白腿小隼（见华中区P170图11）体形差不多，是世界上最小的猛禽之一，数量极少、分布狭窄。二者最明显的区别是红腿小隼比白腿小隼身上多一种红棕色，这是树洞里的红腿小隼飞出的瞬间。

As one of the smallest birds of prey in the world, the red-thighed falconet (*Microhierax caerulescens*) is almost the same as the pied falconet (*Microhierax melanoleucus*, shown in Page 170 Photo11) in size. The most obvious difference between them is that the former has a reddish brown color while the latter has not. This is the moment when the red-thighed falconet flew out of its nest hole in the tree.

❷ 夫妻追逐·云南盈江那邦
Couple's chasing — Nabang, Yingjiang County, Yunnan

线尾燕为一夫一妻制，有细长的突出尾羽是其特征并因此而得名。这是雌雄鸟在相互追逐。

The wire-tailed swallow (*Hirundo smithii*) is monogamous. It is characterized by long and thin prominent tail feathers and so named. In this photo, the male and female birds are chasing each other.

❹ 领地之争·云南盈江铜壁关
Territorial dispute — Tongbi Pass, Yingjiang County, Yunnan

大灰啄木鸟是灰啄木鸟属的一种大型鸟，多栖于树上。而三宝鸟是三宝鸟属下的一种小型攀禽，因侵入了大灰啄木鸟准备用于繁殖的树洞附近而遭到了驱赶却并不示弱。

The great slaty woodpecker (*Mulleripicus pulverulentus*) is a large bird of the genus *Mulleripicus*, which lives on trees. The oriental dollarbird (*Eurystomus orientalis*) in the photo, a small scansores that falls under the genus *Eurystomus*, is being driven away for having invaded into the territory of the great slaty woodpecker where it is about to breed, but seems to be unwilling to leave.

织爱巢·云南盈江那邦
Weaving a nest of love — Nabang, Yingjiang County, Yunnan
黄胸织雀是会用草和长纤维的东西编织巢穴的鸟,为避雨,其巢口向下。这只雄性鸟正在棕榈树下努力地编织精美的爱巢以吸引雌性。
The Baya weaver (*Ploceus philippinus*) weaves its nest with grass and long fiber. To avoid rain, the entrance to the nest mouth is downward. The male bird is trying to create a beautiful nest under the palm tree to attract the female.

↑ 多种多样的两栖爬行动物·华南区
A variety of amphibians and reptiles — SCR

本区的两栖爬行类动物非常丰富，种类繁多，有很多中国特有种，至今还不断有新种被发现。图中依次为：白颌大树蛙（图1）、棕背树蜥（图2）、圆鼻巨蜥（图3）、红蹼树蛙（图4，中国特有种）。

The amphibians and reptiles are rich and diverse in this region, many of them are endemic to China. Up to now, new species are still being discovered from time to time. The animals in the photos are: giant treefrog (*Rhacophorus maximus*, Photo 1); forest garden lizard (*Calotes emma*, Photo 2); common water monitor (*Varanus salvator*, Photo 3); and red webbed tree frog (*Rhacophorus rhodopus*, Photo 4, endemic to China).

潜伏·云南西双版纳 →
Lurking — Xishuangbanna, Yunnan

蟒蛇是世界上巨型蛇类之一，夜行性，生活于热带雨林之中，为捕食猎物往往潜伏无声、伺机而动，总是不断地吐出信子以探测和收集周边的信息。

The python (*Python bivittatus*) is one of the large snakes in the world. It is nocturnal and lives in the tropical rainforest. It often lurks silently and waits for suitable opportunity to attack its prey. Its extruding tongue is particular adept in detecting for preys in places nearby.

← 海南特有鹿·海南大田
A deer endemic to Hainan — Datian, Hainan

坡鹿（俗称海南坡鹿）只生活在海南岛，是我国所有17种鹿中最珍贵的一种。30多年前，坡鹿险些从世界上消失，经过多年的保护工作努力，这个行将灭绝物种又重新恢复了生机。

The Eld's deer (*Panolia siamensis*) only lives in the Hainan Island, which is the most precious of all the 17 kinds of deer found in China. More than 30 years ago, the Eld's deer almost disappeared from the world. After years of conservation efforts, this endangered species has regained its vitality.

❶ 有耳毛的花鼠·云南临沧
Chipmunk with ear hair — Lincang, Yunnan

明纹花松鼠在我国主要分布在云南东南部，耳端有一个黑白毛簇是与其他花鼠的重要区别之一，其分布区也完全不同。

Himalayan striped squirrels (*Tamiops macclellandi*) are mainly distributed in the southeastern Yunnan of China. There is a black-and-white hair cluster at the ear tip, which is one of the important differences between it and other chipmunks that are distributed in totally different regions.

❷ 广布的鼠·云南保山
A ubiquitous squirrel — Baoshan, Yunnan

赤腹松鼠广布于我国南方热带和亚热带森林中，是胃口很大的杂食性动物，吃花算是温柔的，饿起来连树皮都啃以致树木死亡。

Red bellied tree squirrels (*Callosciurus erythraeus*) are widely distributed in the tropical and subtropical forests of the southern China. They are omnivorous animals with a large appetite. Not to mention tender flowers, they would even feed on rough tree barks when hunger strikes, which often causes trees to die.

❸ 罕见大鼠·云南盈江
A rare big squirrel — Yingjiang County, Yunnan

巨松鼠是世界上最大的松鼠，能够生长到1米多长（含尾部），完全生活在树上，其性机警，跳跃力强，非常罕见。

Black giant squirrels (*Ratufa bicolor*) are the largest squirrels in the world. They can grow to more than 1 meter long (including the tail). They only live in trees. Extremely rare, they are alert and particularly good at jumping.

❶ 白头叶猴一家·广西崇左
White headed langur family — Chongzuo, Guangxi

白头叶猴数量稀少,仅分布于广西南部植被繁茂的小范围岩溶地区,栖息范围非常狭窄。这树枝上的一家子好像听到远方有什么响动而一起张望。
White headed langurs (*Trachypithecus leucocephalus*) are scarce in number and only distributed in a small karst area with exuberant vegetation in the south of Guangxi. Their habitat is very narrow. The family on the branch seems to have heard something special in the distance and look toward that direction simultaneously.

❷ 哥俩好·云南西双版纳
Good brothers — Xishuangbanna, Yunnan

北白颊长臂猿为中国、越南、老挝、泰国几国交界边境地区的特有种,分布区非常狭窄。本区的灵长类除分布有北白颊长臂猿和东黑冠长臂猿外,还分布有珍贵的中国特有猿——海南长臂猿。
The north white-cheeked gibbon (*Nomascus leucogenys*) is an endemic species at the borders of China, Vietnam, Laos and Thailand with narrow distribution. In addition to the north white-cheeked gibbon and the eastern black crested gibbon, this region is also home to the rare ape endemic to China, the Hainan gibbon (*Hylobates hainanus*).

舔一舔·云南西双版纳
Licking — Xishuangbanna, Yunnan

蜂猴是完全的树栖动物,喜独自活动,行动特别缓慢,又名"懒猴"。我国分布有3种懒猴,分别是蜂猴、倭蜂猴和间蜂猴,均在云南南部。
As a complete arboreal animal, the Asian slow loris (*Nycticebus bengalensis*) likes to live alone and move slowly. It is also called "lazy monkey". There are three species of loris in China, namely, Asian slow loris, pygmy slow loris (*Nycticebus pygmaeus*) and greater slow loris (*Nycticebus coucang*), all of which are distributed in the south of Yunnan.

从"大象长途北迁"所想到的
Lessons Learnt from the Long-distance Northward Trek of the Wild Herd of Asian Elephants

亚洲象是亚洲现存的最大陆生动物，在我国的历史上，曾广泛分布于华北、华中、华南和西南等地。随着地质年代的变迁及之后人类活动的大大加剧，亚洲象逐渐向南退缩。目前，我国的野生大象仅分布于云南省南部，数量十分稀少且异常珍贵。

新中国成立以来，随着多处自然保护区的建立和野生动物保护力度的不断加大，大象无论是数量上还是种群上都有很大增加，其活动范围也在逐渐扩大。但随之而来的新情况却是，人象冲突的现象越来越频繁，大象损坏庄稼、伤害人畜，甚至致人死亡的事件愈演愈烈。为了寻求解决这些问题的办法，每年全国人民代表大会和政治协商会议提案不少，我就处理过大量的这种提案并多次到实地去考察。

2020年3月，一群野生大象从西双版纳自然保护区开始迁徙，过城串寨一路向北，到2021年6月，居然到达了500千米以外的昆明城郊！野生亚洲象群这样的长途北上，中国首次，世界罕见，一时间成为了国内国际关注的热点。

亚洲象喜群居，每群数头或数十头不等，主食竹笋、嫩枝叶、野芭蕉和棕叶芦（俗称大象草）等。这些食物生长区域的大量减少及破碎化带来大象食物源的匮乏。为了能够摄取赖以生存的食物，大象开始慢慢进入了人类的居住范围，摄取人类种植的农作物，如香蕉、木薯、玉米、水稻和甘蔗等。一开始是因为食物匮乏而偶食，后来慢慢因为这些食物又甜又嫩，大象开始逐渐依赖上了这些食物，于是，大象损坏庄稼、伤害人畜的事件就不断发生，这就更进一步地加剧了人与象的冲突。

为了解决这些矛盾，当地政府和自然保护区采用了建设"大象食堂"的方式来缓解。在保护区外大象活动相对较多的地方，除掉了原有的乔木和灌木甚至农作物，集中种植一些大象喜欢吃的植物，如竹子、芭蕉、玉米、甘蔗等，把大象活动相对控制在一定范围内。这个办法在一定程度上缓解了人象冲突。但是，客观上大象与人的距离不是远了而是近了，人象冲突依然剧烈。

一个需要正视的重要问题是，经过这些年我们保护工作的努力，自然保护区内的森林覆盖率越来越高，热带雨林中原有的林窗、林草结合区域，包括原来刀耕火种方式耕作的区域及不再种庄稼的撂荒地等都随着人口的迁出而逐渐被茂密的森林所覆盖。也就是说，原来这些地方所生长的竹子、野芭蕉、棕叶芦等喜光植物的逐步消失，使大象在自然保护区里可以吃到的食物越来越少了（大象是食草动物，这也是大象"栖息地破坏"的一种表现），逼着大象更多地跑到保护区边缘或保护区外的林草结合区域去取食，其结果是大象都跑出了自然保护区。"大象食堂"固然好，但是像在西双版纳这样的地方，自然保护区已经划建了很大面积，其周边社区人口不少，再占用更多的土地来建设"大象食堂"的潜力有限。

大象能不能更多地待在我们为它们划建的自然保护区里不出来或少出来扰民呢？办法其实还是有的。譬如：我们若将自然保护区里原来大量的林窗、林草结合区域，包括迁出的老百姓不再种庄稼的撂荒地等有计划地定期清理，不让林木去覆盖，尽量让喜光的草本植物生长，修复成原来大象的栖息地，尽可能让大象留在自然保护区里不出去或少出去扰民，人象冲突的现象就能够大大减缓。我们应该充分认识到，越来越茂密的森林并不是大象最佳的栖息地，我们建立这些自然保护区保护的主要对象是中国非常珍贵的大象，我们为什么不能为了这个主要保护对象而改变我们自然保护区管理"一刀切"的陈腐理念呢？

可惜这个办法与现行的《中华人民共和国自然保护区管理条例》（以下简称《条例》）是冲突的（目前有关部门正在为修改本条例广泛征求意见）。现行条例规定，自然保护区里（尤其是核心区、缓冲区）是"一草一木不能动"的。这种自然保护区管理规定与保护的主要对象完全相悖的例子在中国的实际工作中还有不少。例如在朱鹮自然保护区，朱鹮靠农民耕种的水田里的泥鳅、田螺等维持生活，在村庄周边的大树上栖息，如果按现行条例执行，在自然保护区里"一草一木不能动"，还要把人都迁走，那么朱鹮也就不复存在了。保护野生动物最根本、最有效的办法是栖息地的保护与修复，自然修复和人工促进修复都是栖息地保护工作应有的选择。尤其是一些与人类关系密切的野生动物，如亚洲象、蒙原羚、朱鹮等，必须要采取有针对性的措施来分类指导、分区施策。

另外，还有一个环境容量问题。我国由于保护得力，亚洲象已从原来的150头左右翻番到300头左右。那么，在西双版纳、在云南甚至在中国，我们到底能够提供多少大象的栖息地让大象与人和谐共处？大象数量有多少是适宜的？这是我们当前以及今后都必须要解决的大课题。在我国自然保护力度越来越大的背景下，亚洲象并不是第一个也不是最后一个给我们提出这个考题的野生动物（还有如东北的虎豹、贺兰山等地的岩羊、青藏高原的部分有蹄类动物，甚至很多地方的野猪、猴子等），这类现象今后只会越来越多。

"师法自然"是人与自然到底应该如何相处的哲学回答，目前我们需要做的是：乘"建立以国家公园为主体的自然保护地体系"的改革东风，修改"一刀切"的现行《条例》，对于中国的自然保护地实行"分级分类分区"管理（本人已呼吁多年），尤其是对于保护的主要对象是野生动物的自然保护地不是封闭起来就行，而必须进行分类指导、分区施策，在自然保护地内应该设置不同的有针对性的功能分区，为特定保护的野生动物提供适宜的、面积合适的栖息地，再用"一区一法"的办法（包括按相同类型归类的多区一法和特殊保护地的一区一法）固定下来。只有这样，我们的自然保护地才能够确确实实解决许多前进中暴露出来的问题，才能做到人与自然和谐相处，才能真正实现人与野生动物同为"生命共同体"的美好愿景。

The Asian elephant (*Elephas maximus*) is the largest terrestrial animal in Asia that used to be widely distributed in North China, Central China, South China and South-west China. With the evolution of geological ages and the increasing infringement of human activities, the Asian elephant gradually retreated to an ever shrinking niche in the southern regions of the country. At present, wild elephants in China are only distributed in the southern parts of the Yunnan Province, making them extremely rare and precious.

As more nature reserves have been established since the founding of the People's Republic of China, and thanks to the increasingly stronger commitments and efforts made in wildlife protection, both the total population of elephants and the number of their herds have seen notable growth, and the roaming ranges of the elephants have also expanded steadily. But as a consequence, the human-elephant conflict is becoming more and more frequent, with increasing damages brought by elephants to crops, domestic animals and people, and even loss of human lives. In order to protect the elephants without inflicting elephant-caused economic loss and casualties on the local people, the local government has taken many measures to alleviate these conflicts. Numerous motions and proposals have been put forward each year during the National People's Congress and the Chinese People's Political Consultative Conference sessions to tackle this problem. I myself have frequently taken part in the deliberations concerning these proposals and motions and have been dispatched on field trips to look into the feasibilities of putting them into practices.

In March 2020, a herd of wild elephants left the Xishuangbanna Nature Reserve and embarked on their north-bound trek. By June 2021, they had already covered over 500 km and arrived at the outskirts of Kunming, the capital city of Yunnan Province. This long trek of wild elephants herd to the north, which is not only unprecedented in China, but also rarely seen throughout the world, has caught extensive media coverage in both China and the globe.

Asian elephants like to live in groups that range in size from less than ten to dozens in each group. They mainly eat bamboo shoots, tender leaves, wild plantains and tiger grass (commonly known as elephant weed). Ever shrinking areas that are suitable for these plants to grow, together with serious fragmentation of their habitats, has led in consequence to insufficient food supply for elephants. In order to obtain the food that they depend on, elephants have slowly entered areas inhabited by human beings, where farmer-grown crops like bananas, cassava, corn, rice paddy and sugarcane are available. In the beginning, elephants only ate such crops occasionally when their normal food is in short supply. But gradually, they realized that such foods are so tasty and tender that they became dependent on them. As a result, the incidents of elephants damaging crops and hurting people as well as domestic animals occurred with higher frequencies, which further intensified the conflicts between people and elephants.

To address this problem, local governments and corresponding nature reserve managers have come up with an idea which proposes that special places would be reserved and allocated as the "elephant canteens". In elephant-haunted places beyond the boundaries of nature reserves, trees and shrubs that naturally grown there and

← 雨林之王·云南西双版纳
The king of rainforest — Xishuangbanna, Yunnan

图为雨林中的亚洲象象王。象牙给大象带来了生活取食方便的同时，更给它们带来了巨大的伤害，人们因为要获取象牙而杀害大象。为了呼吁人们关注身处迫切困境的非洲象和亚洲象，每年8月12日被确立为"世界大象日"。

Standing in this photo is the king of the Asian elephant (*Elephas maximus*) that inhabits in the rainforests. Tusks of elephants are both a bless and a curse at the same time, for while they make up a useful tool for the elephants to live and eat, they also bring great harm to the animal. People kill elephants because they want to get ivory from the elephants. In order to raise people's awareness on the endangered situation that African elephants and Asian elephants are faced with, August 12 is designated as "World Elephant Day".

喜水的亚洲象·
云南西双版纳
Water-loving Asian elephants —
Xishuangbanna, Yunnan

亚洲象是草食性动物，主要栖息于热带雨林、季雨林及林间生长有大量草类植物的沟谷、山坡、平缓地等区域。大象皮厚、皱多、毛发少，易受蚊虫叮咬和生皮肤病，需要经常洗澡或泥浴，而且每天还需要摄入大量的水分，所以足够的草类植物和水在大象生活中是必不可少的。这些生存条件的要求，确定了大象栖息地的性质。

As vegetarian animals, Asian elephants primarily inhabit in tropical rainforests, monsoon rainforests, and in other types of forests that grow in deep valleys, hill slopes and flat lands where abundant herbal plants and grasses are available. In addition to huge daily consumption, water are also needed for taking bathes (sometimes with dirt), since elephants' hair-scarce and heavily-wrinkled thick skin is highly susceptible to mosquito or insect bites, bacterial infections and other skin diseases. As a result, abundant water and grass supply makes up a defining feature of the habitats that meet the needs of elephants.

even farmland crops would be deliberated removed and replaced with heliophiles that elephants like to eat, such as bamboo, hardy banana, corn, sugarcane and *etc.*, thus limiting the movement of elephants within the scopes of the designated areas. To a certain extent, this method does play a role in alleviating the human-elephant conflicts. However, the employment of such method has, instead of keeping the elephants far away from human settlements, in fact further narrowed down the distance between the two parts. As a result, serious conflicts still persist.

An important factor that we should face up to is that, through our efforts over the years, the forests cover in the nature reserves has been notably improved. The windows or glades and the forest-grass interfacing borders previously existed in the tropical rainforests, together with areas where such primitive methods as slash-and-burn farming were adopted as the primary means of land-using, as well as the abandoned farmlands, have so far all been overgrown with dense forests as people previously dwelling here moved away. An unexpected consequence is that heliophiles such as bamboos, hardy bananas and tiger grasses that formerly flourished in such places have also gradually disappeared, making food supply for elephants within the nature reserves less abundant (because the elephant is a herbivorous animal, this is yet a kind of damage inflicted upon the habitats of elephants). As a result, the elephants have no choice but to go beyond the boundaries of nature reserves and to search for food in places where dense forests border on grasslands. The idea of setting up "elephant canteens" itself is good, but in places like Xishuangbanna, nature reserves have already taken up a huge portion of lands available there and, given the existence of local communities whose populations are by no means small, it is obviously not feasible for local people to give up more of their farm lands for the sake of setting up more "elephant canteens".

Is it possible for us to limit the elephants within the nature reserves that are specially set up for their protection, and hence cutting down their potential harassments to local people? Yes, as a matter of fact, it is. Take for instance, if we regularly clear up trees to restore the forests windows or glades and the forest-grass interfacing borders that had previously existed in the tropical forests, clear up the abandoned farmlands that fall within the nature reserves, so as to make sure that such areas will not be covered with dense forests, hence providing a favorable amenity for the growth of herbaceous heliophiles that are favored by elephants and in turn retoring them into "elephant habitats", the elephants will be never or far less likely to go out of the nature reserves to disturb the people. The human-elephant conflict will naturally be mitigated to a great extent. We must be fully aware that forests that are getting increasingly denser do not make up optimal habitats for elephants. Given that the central goal of setting up these nature reserves is to protect the precious elephants in China, why don't we just discard the stereotyped and outdated "one-size-fits-all" solution that is indiscriminately applied to all types of nature reserves?

Unfortunately, this proposal does not conform to the *Regulations of the People's Republic of China on the Administration of Nature Reserves* (hereinafter referred to as the *Regulations*) that is currently in effect (at the present, the legislative authorities are soliciting comments and suggestions concerning the amendment of the *Regulations*), which stipulates that the removal of anything, even a single tree or grass within the nature reserves (especially in the core zone and the buffer zone) shall not be allowed. There are many cases in China where such regulations turn out to be counter-productive to the main goals of protective efforts. For example, in the crested ibis nature reserves, the crested ibis lives on loaches and snails in the farmers' cold-water crop field and perches on the big trees around the village. If measures were implemented mechanically in accordance with the current *Regulations* — "the removal of anything, even a single tree or grass, within nature reserves shall not be allowed" — and if all farmers were repositioned away from the nature reserves, they would, rather than playing any constructive roles, only end up with poor results in the protection of the crested ibis. The essential and most effective way for wildlife conservation is habitat conservation and restoration through the application of both nature-based and man-aided restorative endeavors. This is particularly true with regard to the conservation of wildlife species that bear relatively closer affinity to human beings, such as Asian elephants, Mongolian gazelles, crested ibis and *etc*., whose effective conservation calls for tailor-made policies that take the demands of the specific species as well as the actual situations of the region at issue into due considerations.

Another key consideration is the loading capacity of the site. Thanks to the effective conservational endeavors that have so far been carried out in China, the total population of Asian elephants has boosted from around 150 to 300, marking an increase by 100%. Questions that follow naturally would be: What is the suitable scope of habitats that can be provided in Xishuangbanna, in Yunnan, or in the whole country of China? What is the proper elephant population that would guarantee harmonious co-existence between human beings and the animal? These are all questions that urgently need to be answered at the present, and they will remain so for a fairly long period of time in the future. Under the current context where China is increasingly strengthening its nature conservation efforts, Asian elephants are not the first wildlife species, and by no means will be the last one either, to pose these challenging questions to us. Such challenges that we will have to tackle with will only become more frequent. (Telling examples include tigers and leopards in the north-east, blue sheep in the Helan Mountains, certain ungulate species inhabiting the Qinghai-Tibet Plateau, and even wild boars and monkeys in some parts of the country.)

"Drawing on the laws of nature" is a philosophy in which resides the ultimate answer to the question as to "how we human beings can co-exist harmoniously with our surrounding environment". What we need to do at present is to, taking advantage of the on-going reform that aims to put in place "a natural protected area system highlighting the central role that national parks play", eliminate some of the "one-size-fits-all" approaches stipulated in the *Regulations* that is currently in effect through making due amendments to the *Regulations*. Discretionary approaches that take the specific situation of each protected area

into account (for which I have been championing over the years) should be adopted. Especially for the protected areas whose main target is wildlife, it is far from enough to merely put them under closure. Instead, tailor-designed policies that take into consideration the demands of the specific species and places at issue must be formulated, so as to break down the protected areas into different functional zones that meet the demands of the protected wildlife species for suitable habitats in terms of both their sizes and environment. On basis of this, case-by-case approaches (including adopting the same approach for multiple sites that share similar purposes and features, or a uniquely-designed approach for protected area that serve a certain unique purpose) should be adopted. Only in this way can the many problems that we have encountered in the management of our nature reserves be properly addressed, can harmonious coexistence between human and nature be achieved, and can the vision of human and wildlife as a "community of life" be realized.

肇事象群出现了·
云南普洱江城
The perpetrating elephants caught red-handed —
Jiangcheng of Pu'er, Yunnan

这群野生亚洲象是20世纪90年代起逐渐远离西双版纳自然保护区进入几十千米外的江城县的，现在共有19头，常常出入城镇、农村，进入老百姓家，翻箱倒柜、践踏庄稼、伤害人畜。迄今为止，这群肇事象群已经背负20多条人命案，伤者无数。我和云南亚洲象监测中心的人员一直等到傍晚时节，这群野象终于从森林里出来了，它们肆意在香蕉林里挥霍，到茶园里践踏，并逐渐向我们走来。

Since the 1990s, this herd of wild Asian elephants have gradually moved away from Xishuangbanna Nature Reserve and invaded into Jiangcheng County, which is situated dozens of kilometers away. Now there are 19 elephants in the group. They often invade towns, villages and people's homes, to wreak havoc, causing serious damages to farm crops, domestic animals, and sometimes even hurting human beings. So far, this group of elephants has been responsible for the death over 20 human-lives and countless injuries. The staff of Yunnan Asian Elephant Monitoring Center and I waited for a long time until the wild elephants finally came out of the forest in the evening. They had wreaked havoc in the banana groves and the tea farm, before moving forward in our direction.

物种名录
List of Species

为了便于读者进一步了解物种情况和查阅，此部分列出了在本书中出现的物种（按物种中文名拼音的首字母排序）及其中文名、拉丁学名、保护级别、濒危程度等，还标上了该物种在书中出现的页码。

注：为保护野生动物，维护生态平衡，我国颁布了《中华人民共和国野生动物保护法》，其中规定国家对珍贵、濒危的野生动物实行重点保护，并根据物种的珍贵濒危程度和管理严格要求依次分为一级保护和二级保护（简写为国家一级和国家二级）。本书物种的国家保护级别依据《国家重点保护野生动物名录》（2021年2月1日公布），分别简写为CHINA I 和CHINA II 。

世界自然保护联盟（简称IUCN）是目前世界上最大的、最重要的世界性保护联盟。IUCN编制的《世界自然保护联盟濒危物种红色名录》是被广泛接受和使用的受威胁物种分级标准体系。该组织每年评估数以千计物种的绝种风险，将物种编入9个不同的保护级别：依次为灭绝（EX）、野外灭绝（EW）、极危（CR）、濒危（EN）、易危（VU）、近危（NT）、无危（LC）、数据缺乏（DD）和未予评估（NE）。本书物种的IUCN保护级别依据2020年1月15日后查询的《世界自然保护联盟濒危物种红色名录》（IUCN）官方网站，分别简写为IUCN EX、IUCN EW、IUCN CR、IUCN EN、IUCN VU、IUCN NT、IUCN LC、IUCN DD和IUCN NE。

《濒危野生动植物种国际贸易公约》（简称CITES）是全球缔约国之间为了保护野生动植物物种不至于由于国际贸易而遭到过度开发利用而进行的国际合作。CITES将受管理的野生动植物物种按照其物种状况及其受贸易影响的严重程度依次列为附录 I 、附录 II 和附录 III 名单。本书物种的CITES保护级别依据2019年11月26日公布的《濒危野生动植物种国际贸易公约》，分别简写为CITES I 、CITES II 和CITES III 。

本书中鸟类、兽类、爬行动物名录主要参考以下文献。

郑光美. 中国鸟类分类与分布名录[M]. 2版. 北京: 科学出版社, 2017.

蒋志刚, 刘少英, 吴毅, 等. 中国哺乳动物多样性(第2版)[J]. 生物多样性, 2017, 25 (8): 886–895.

蔡波, 王跃招, 陈跃英, 等. 中国爬行纲动物分类厘定[J]. 生物多样性, 2015, 23(3): 365-382.

For the purpose of providing more detailed information about the species and for the sake of easier reference for the readers, this section lists the species that are covered in the book (sorted by the first letter of the pin yin of Chinese name of the species in alphabetical order), including their Chinese names, Latin names, protection levels, to what extent the species is endangered and *etc.*. The page number in which the species appears in the book is also indicated.

Notes: In order to protect wildlife and maintain ecological balance, China has promulgated the *Law of the People's Republic of China on Wild Animals Protection,* which stipulate that China implements prioritized protection for rare and endangered wild animals, and that, depending on the extent to which that are rare or endangered as well as on the strictness of protection, the wild animals under protection have been divided into Class I and Class II Key Protected Wild Animals of National Significance respectively (abbreviated as the national Class I and national Class II). The national protection level of species in this book is based on the *List of Key Protected Wild Animals of National Significance* (issued on February 1, 2021), abbreviated as CHINA I and CHINA II respectively.

The International Union for Conservation of Nature (IUCN) is the world's largest and most important conservation union in the world. The *IUCN Red List of Endangered Species* is a widely accepted and used classification system for threatened species. Each year, the organization evaluates the extinction risk of thousands of species, and classifies them into nine different protection levels: Extinct (EX), Extinct in the Wild (EW), Critically Endangered (CR), Endangered (EN), Vulnerable (VU), Near Threatened (NT), Least Concern (LC),

Data Deficient (DD) and Not Evaluated (NE). The IUCN protection level of the species covered in this book is based on the information retrieved from the official website of the *IUCN Red List of Endangered Species*, on and after January 15, 2020, abbreviated as IUCN EX, IUCN EW, IUCN CR, IUCN EN, IUCN VU, IUCN NT, IUCN LC, IUCN DD and IUCN NE respectively.

The *Convention on International Trade in Endangered Species of Wild Fauna and Flora* (CITES) promotes international cooperation between the contracting parties among the world to protect the species of wild fauna and flora from over exploitation and utilization due to international trade. CITES lists the managed wildlife species in CITES Appendix Ⅰ, CITES Appendix Ⅱ and CITES Appendix Ⅲ according to their species status and the severity they are affected in trade. The CITES protection level of the species in this book is based on the *Convention on International Trade in Endangered Species of Wild Fauna and Flora* published on November 26, 2019, abbreviated as CITES Ⅰ, CITES Ⅱ and CITES Ⅲ respectively.

The author has consulted the following main literatures on the names of the birds, beasts and reptiles covered in this book.

Zheng Guangmei. Catalogue of Bird Classification and Distribution in China [M]. 3rd Edition. Beijing: Science Press, 2017.

Jiang Zhigang, Liu Shaoying, Wu Yi, et al. Mammal Diversity in China (2nd Edition) [J]. Biodiversity, 2017, 25 (8): 886-895.

Cai Bo, Wang Yuezhao, Chen Yueying, et al. Taxonomic Determination of Reptiles in China[J]. Biodiversity, 2015, 23 (3): 365-382.

阿尔泰盘羊 ·············· P97
Ovis ammon
CHINA Ⅱ, CITES Ⅱ, IUCN NT

白腹锦鸡 ·············· P137
Chrysolophus amherstiae
CHINA Ⅱ, IUCN LC

白腹鹞 ·············· P43
Circus spilonotus
CHINA Ⅱ, CITES Ⅱ, IUCN LC

白冠长尾雉 ·············· P78
Syrmaticus reevesii
CHINA Ⅰ, CITES Ⅱ, IUCN VU

白颌大树蛙 ·············· P204
Rhacophorus maximus
IUCN LC

白鹤 ·············· P40, 168
Grus leucogeranus
CHINA Ⅰ, CITES Ⅰ, IUCN CR

白颊山鹧鸪 ·············· P139
Arborophila atrogularis
IUCN NT

白琵鹭 ·············· P70
Platalea leucorodia
CHINA Ⅱ, IUCN EN

白头叶猴 ·············· P208
Trachypithecus leucocephalus
CHINA Ⅰ, CITES Ⅱ, IUCN CR

白腿小隼 ·············· P171
Microhierax melanoleucos
CHINA Ⅱ, CITES Ⅱ, IUCN LC

 白臀鹿 .... P121 *Cervus wallichii macneilli* CHINA Ⅰ, IUCN LC	 白尾海雕 .... P46 *Haliaeetus albicilla* CHINA Ⅰ, CITES Ⅰ, IUCN LC	 白尾梢虹雉 .... P138 *Lophophorus sclateri* CHINA Ⅰ, CITES Ⅰ, IUCN VU	 白鹇 .... P137 *Lophura nycthemera* CHINA Ⅱ, IUCN LC	 白胸翡翠 .... P148 *Halcyon smyrnensis* CHINA Ⅱ	

白胸苦恶鸟 P148
Amaurornis phoenicurus
IUCN LC

白腰杓鹬 P71
Numenius arquata
IUCN NT

白枕鹤 P39
Grus vipio
CHINA Ⅰ, CITES Ⅰ, IUCN VU

斑喉希鹛 P171
Minla strigula
IUCN LC

斑尾塍鹬 P68
Limosa lapponica
IUCN NT

北白颊长臂猿 P208
Nomascus leucogenys
CHINA Ⅰ, CITES Ⅰ, IUCN CR

北山羊 P96
Capra sibirica
CHINA Ⅱ, IUCN LC

长江江豚 P174, 176
Neophocaena asiaeorientalis asiaeorientalis
CHINA Ⅰ, CITES Ⅰ, IUCN VU

长尾林鸮 P56, 102
Strix uralensis
CHINA Ⅱ, CITES Ⅱ, IUCN LC

橙腹叶鹎 P171
Chloropsis hardwickii
IUCN LC

赤腹松鼠 P207
Callosciurus erythraeus
IUCN LC

赤狐 P95
Vulpes vulpes
IUCN LC

赤颈鹤 P199
Grus antigone
CHINA Ⅰ, CITES Ⅱ, IUCN VU

川金丝猴 P8, 140
Rhinopithecus roxellana
CHINA Ⅰ, CITES Ⅰ, IUCN EN

穿山甲 P146
Manis pentadactyla
CHINA Ⅰ, CITES Ⅰ, IUCN CR

大白鹭 P76, 81
Ardea alba
IUCN LC

大鸨 P74
Otis tarda
CHINA Ⅰ, CITES Ⅱ, IUCN VU

大灰啄木鸟 P201
Mulleripicus pulverulentus
CHINA Ⅱ, IUCN VU

大鵟 P103
Buteo hemilasius
CHINA Ⅱ, CITES Ⅱ, IUCN LC

大麻鳽 P71
Botaurus stellaris
IUCN LC

大鲵 P167
Andrias davidianus
CHINA Ⅱ, CITES Ⅰ, IUCN CR

大拟啄木鸟 P148
Psilopogon virens
IUCN LC

大天鹅 P80
Cygnus cygnus
CHINA Ⅱ, IUCN LC

大熊猫 P4, 131, 132, 135
Ailuropoda melanoleuca
CHINA Ⅰ, CITES Ⅰ, IUCN VU

大嘴乌鸦 P122
Corvus macrorhynchos
IUCN LC

 玳瑁 ......P194 *Eretmochelys imbricata* CHINA I , CITES I , IUCN CR	 丹顶鹤 ......P6,38 *Grus japonensis* CHINA I , CITES I , IUCN EN	 滇金丝猴 ......P142 *Rhinopithecus bieti* CHINA I , CITES I , IUCN EN	 雕鸮 ......P45 *Bubo bubo* CHINA II , CITES II , IUCN LC	 东北虎 ......P35,37 *Panthera tigris altaica* CHINA I , CITES I , IUCN EN
 鹅喉羚 ......P98 *Gazella subgutturosa* CHINA II , IUCN VU	 翻车鱼 ......P194 *Mola mola* IUCN VU	 反嘴鹬 ......P71 *Recurvirostra avosetta* IUCN LC	 菲氏叶猴 ......P143 *Trachypithecus phayrei* CHINA I , CITES II , IUCN EN	 蜂猴 ......P209 *Nycticebus bengalensis* CHINA I , CITES I , IUCN VU
 凤头麦鸡 ......P71 *Vanellus vanellus* IUCN LC	 凤头䴙䴘 ......P43,77 *Podiceps cristatus* IUCN LC	 高鼻羚羊 ......P98 *Saiga tatarica* CHINA I , CITES II , IUCN CR	 高黎贡白眉长臂猿 ......P144 *Hoolock tianxing* CHINA I , CITES I , IUCN EN	 高山兀鹫 ......P122 *Gyps himalayensis* CHINA II , CITES II , IUCN NT
 冠斑犀鸟 ......P186 *Anthracoceros albirostris* CHINA I , CITES II , IUCN LC	 河狸 ......P94 *Castor fiber* CHINA I , IUCN LC	 河燕鸥 ......P201 *Sterna aurantia* CHINA I , IUCN NT	 褐马鸡 ......P78 *Crossoptilon mantchuricum* CHINA I , CITES I , IUCN VU	 黑顶蛙口夜鹰 ......P200 *Batrachostomus hodgsoni* CHINA II , IUCN LC
 黑颈长尾雉 ......P136 *Syrmaticus humiae* CHINA I , CITES I , IUCN NT	 黑颈鹤 ......P150 *Grus nigricollis* CHINA I , IUCN LC	 黑琴鸡 ......P105 *Lyrurus tetrix* CHINA I , IUCN LC	 黑头蜡嘴雀东北亚种 ......P49 *Eophona personata magnirostris* IUCN LC	 黑头奇鹛 ......P171 *Heterophasia melanoleuca* IUCN LC
 黑尾鸥 ......P72 *Larus crassirostris* IUCN LC	 黑胸太阳鸟 ......P171 *Aethopyga saturata* IUCN LC	 黑熊 ......P53 *Ursus thibetanus* CHINA II , CITES I , IUCN VU	 黑嘴鸥 ......P71 *Saundersilarus saundersi* CHINA I , IUCN VU	 红白鼯鼠 ......P146 *Petaurista alborufus* IUCN LC

红腹角雉 ·········· P139
Tragopan temminckii
CHINA II , IUCN LC

红腹锦鸡 ·········· P78
Chrysolophus pictus
CHINA II , CITES II , IUCN LC

红喉山鹧鸪 ·········· P199
Arborophila rufogularis
CHINA II , IUCN LC

红脚鲣鸟 ·········· P199
Sula sula
CHINA II , IUCN LC

红蹼树蛙 ·········· P204
Rhacophorus rhodopus
IUCN LC

红腿小隼 ·········· P201
Microhierax caerulescens
CHINA II , CITES II , IUCN LC

红尾水鸲 ·········· P171
Rhyacornis fuliginosa
IUCN LC

红原鸡 ·········· P199
Gallus gallus
CHINA II , IUCN LC

红嘴鸥 ·········· P148
Chroicocephalus ridibundus
IUCN LC

鸿雁 ·········· P74
Anser cygnoid
CHINA II , IUCN VU

胡兀鹫 ·········· P122
Gypaetus barbatus
CHINA I , CITES II , IUCN NT

花冠皱盔犀鸟 ·········· P187
Rhyticeros undulatus
CHINA I , CITES II , IUCN VU

花龟 ·········· P167
Mauremys sinensis
CHINA II , CITES III , IUCN EN

华南虎 ·········· P165
Panthera tigris amoyensis
CHINA I , CITES I , IUCN CR

环颈雉 ·········· P79
Phasianus colchicus
IUCN LC

黄颈凤鹛 ·········· P148
Yuhina flavicollis
IUCN LC

黄腿渔鸮 ·········· P171
Ketupa flavipes
CHINA II , CITES II , IUCN LC

黄胸织雀 ·········· P202
Ploceus philippinus
CITES II , IUCN LC

黄爪隼 ·········· P49
Falco naumanni
IUCN LC

灰背伯劳 ·········· P171
Lanius tephronotus
IUCN LC

灰翅浮鸥 ·········· P77
Chlidonias hybrida
IUCN LC

灰鹤 ·········· P42
Grus grus
CHINA II , CITES II , IUCN LC

灰孔雀雉 ·········· P198
Polyplectron bicalcaratum
CHINA I , CITES II , IUCN LC

金背啄木鸟 ·········· P148
Dinopium javanense
IUCN LC

金雕 ·········· P103
Aquila chrysaetos
CHINA I , CITES II , IUCN LC

金眶鸻 ·········· P71
Charadrius dubius
IUCN LC

鲸鲨 ·········· P197
Rhincodon typus
CHINA II , CITES II , IUCN EN

巨松鼠 ·········· P207
Ratufa bicolor
CHINA II , CITES II , IUCN NT

蓝翅希鹛 ·········· P148
Siva cyanouroptera
IUCN LC

蓝冠噪鹛 ·········· P171
Garrulax courtoisi
CHINA I , IUCN CR

 蓝喉蜂虎 ·········· P171 Merops viridis CHINA II , IUCN LC	 蓝喉太阳鸟 ·········· P172 Aethopyga gouldiae IUCN LC	 狼 ·········· P94 Canis lupus CHINA II , CITES II , IUCN LC	 栗额斑翅鹛 ·········· P148 Actinodura egertoni IUCN LC	 栗头蜂虎 ·········· P171 Merops leschenaulti CHINA II , IUCN LC
 蛎鹬 ·········· P71 Haematopus ostralegus IUCN NT	 林鹬 ·········· P71 Tringa glareola IUCN LC	 绿海龟 ·········· P192,194 Chelonia mydas CHINA I , CITES I , IUCN EN	 绿孔雀 ·········· P184 Pavo muticus CHINA I , CITES II , IUCN EN)	 马鹿 ·········· P54 Cervus canadensis CHINA II , IUCN LC
 蟒蛇 ·········· P205 Python bivittatus CHINA II , CITES II , IUCN VU	 毛腿沙鸡 ·········· P104 Syrrhaptes paradoxus IUCN LC	 梅花鹿 ·········· P53 Cervus nippon CHINA I , IUCN LC	 蒙古野驴 ·········· P98 Equus hemionus CHINA I , CITES I , IUCN NT	 蒙原羚 ·········· P99,107 Procapra gutturosa CHINA I , IUCN LC
 麋鹿 ·········· P63,65 Elaphurus davidianus CHINA I , IUCN EW	 庙岛蝮 ·········· P75 Gloydius lijianlii IUCN LC	 明纹花松鼠 ·········· P207 Tamiops macclellandi IUCN LC	 牛背鹭 ·········· P65 Bubulcus ibis IUCN LC	 怒江金丝猴 ·········· P141 Rhinopithecus strykeri CHINA I , CITES I , IUCN CR
 狍 ·········· P51 Capreolus pygargus IUCN LC	 坡鹿 ·········· P206 Panolia siamensis CHINA I , CITES I , IUCN EN	 普氏野马 ·········· P87,P89 Equus ferus CHINA I , CITES I , IUCN EN	 普氏原羚 ···· P107,114,115 Procapra przewalskii CHINA I , IUCN EN	 普通海鸥 ·········· P71 Larus canus IUCN LC
普通䴓 ·········· P49 Sitta europaea IUCN LC	 黔金丝猴 ·········· P166 Rhinopithecus brelichi CHINA I , CITES I , IUCN EN	 钳嘴鹳 ·········· P149 Anastomus oscitans IUCN LC	 青海湖裸鲤 ·········· P116,117 Gymnocypris przewalskii	 青脚鹬 ·········· P71 Tringa nebularia IUCN LC

221

青头潜鸭 P71
Aythya baeri
CHINA I, IUCN CR

热带点斑原海豚 P193
Stenella attenuata
CHINA II, CITES II, IUCN LC

三宝鸟 P201
Eurystomus orientalis
IUCN LC

三趾翠鸟 P199
Ceyx erithaca
IUCN LC

沙蜥 P49,104
Phrynocephalus sp.
IUCN LC

猞猁 P94
Lynx lynx
CHINA II, CITES II, IUCN LC

石鸡 P104
Alectoris chukar
IUCN LC

双角犀鸟 P188
Buceros bicornis
CHINA I, CITES I, IUCN VU

水獭 P159
Lutra lutra
CHINA II, CITES II, IUCN NT

水雉 P170
Hydrophasianus chirurgus
CHINA II, IUCN LC

四川羚牛 P147
Budorcas tibetanus
CHINA I, CITES II, IUCN VU

蓑羽鹤 P42
Grus virgo
CHINA II, CITES II, IUCN LC

塔里木马鹿 P98
Cervus yarkandensis
CHINA I, IUCN LC

台湾鬣羚 P191
Capricornis swinhoei
CHINA I, IUCN LC

台湾梅花鹿 P190
Cervus nippon taiouanus
CHINA I, IUCN LC

秃鹫 P123
Aegypius monachus
CHINA I, CITES II, IUCN NT

乌林鸮 P45
Strix nebulosa
CHINA II, CITES II, IUCN LC

西黑冠长臂猿 P145
Nomascus concolor
CHINA I, CITES I, IUCN CR

西太平洋斑海豹 P10,66,67
Phoca largha
CHINA I, IUCN LC

西藏马鹿 P121
Cervus wallichii
CHINA I, IUCN NT

线尾燕 P201
Hirundo smithii
IUCN LC

小麂 P167
Muntiacus reevesi
IUCN LC

小天鹅 P100
Cygnus columbianus
CHINA II, IUCN LC

小熊猫 P152
Ailurus fulgens
CHINA II, CITES I, IUCN EN

熊猴 P145
Macaca assamensis
CHINA II, CITES II, IUCN NT

雪豹 P120
Panthera uncia
CHINA I, CITES I, IUCN VU

雪鸮 P44
Bubo scandiacus
CHINA II, CITES II, IUCN VU

血雀 P148
Carpodacus sipahi
IUCN LC

血雉 P139
Ithaginis cruentus
CHINA II, CITES II, IUCN LC

驯鹿 P50
Rangifer tarandus
CITES I, IUCN VU

亚洲象 ······ P210,213,214
Elephas maximus
CHINA I , CITES I , IUCN EN

岩鹭 ······ P199
Egretta sacra
CHINA II , IUCN LC

岩松鼠 ······ P50
Sciurotamias davidianus
IUCN LC

岩羊 ······ P97
Pseudois nayaur
CHINA II , IUCN LC

眼纹噪鹛 ······ P171
Garrulax ocellatus
CHINA II , IUCN LC

扬子鳄 ······ P162,163
Alligator sinensis
CHINA I , CITES I , IUCN CR

野骆驼 ······ P92
Camelus ferus
CHINA I , IUCN CR

野牦牛 ······ P113
Bos mutus
CHINA I , CITES I , IUCN VU

遗鸥 ······ P90,91
Ichthyaetus relictus
CHINA I , CITES I , IUCN VU

银耳相思鸟 ······ P148
Leiothrix argentauris
CHINA II , CITES II , IUCN LC

银喉长尾山雀 ······ P49
Aegithalos glaucogularis
IUCN LC

圆鼻巨蜥 ······ P204
Varanus salvator
CHINA I , CITES II , IUCN LC

藏羚 ······ P124,125
Pantholops hodgsonii
CHINA I , CITES I , IUCN NT

藏酋猴 ······ P145
Macaca thibetana
CHINA II , CITES II , IUCN NT

藏鼠兔 ······ P121
Ochotona thibetana
IUCN LC

藏雪鸡 ······ P104
Tetraogallus tibetanus
CHINA II , CITES I , IUCN LC

藏野驴 ······ P118
Equus kiang
CHINA I , CITES II , IUCN LC

藏原羚 ······ P121
Procapra picticaudata
CHINA II , IUCN NT

中国鲎 ······ P194
Tachypleus tridentatus
CHINA II , IUCN EN

中华白海豚 ······ P196,197
Sousa chinensis
CHINA I , CITES I , IUCN VU

中华秋沙鸭 ······ P48
Mergus squamatus
CHINA I , IUCN EN

中华鲟 ······ P177
Acipenser sinensis
CHINA I , CITES II , IUCN CR

朱鹮 ······ P160,161
Nipponia nippon
CHINA I , CITES I , IUCN EN

紫貂 ······ P50
Martes zibellina
CHINA I , IUCN LC

紫水鸡 ······ P148
Porphyrio porphyrio
CHINA II , IUCN LC

紫啸鸫 ······ P171
Myophonus caeruleus
IUCN LC

棕背树蜥 ······ P204
Calotes emma

棕头鸥 ······ P117
Chroicocephalus brunnicephalus
IUCN LC

棕胸竹鸡 ······ P199
Bambusicola fytchii
IUCN LC

棕熊 ······ P52
Ursus arctos
CHINA II , CITES I , IUCN LC

后记

2020年年初，一场百年未遇的重大疫情——新型冠状病毒席卷了全球。中国是疫情暴发最早的国家，也是世界上疫情控制得最早、最好的国家之一。目前，全国已经基本控制住了疫情，全面复工复产复课，生活生产都进入了正常轨道。与此同时，中国还将大量的抗疫物资、经验和疫苗分享给了世界很多正在抗疫紧要关头的国家。这些都是中国政府采取了强有力的手段和及时正确措施的结果，体现了中国的大国担当。

在疫情发展过程中，2020年2月24日全国人民代表大会常务委员会颁布了《关于全面禁止非法野生动物交易、革除滥食野生动物陋习、切实保障人民群众生命健康安全的决定》。这个决定聚焦滥食野生动物的突出问题，目的就是要在相关法律修改之前，全面禁止食用野生动物，严厉打击非法野生动物交易，为维护公共卫生安全和生态安全，保障人民群众生命健康安全提供有力的立法保障。它的贯彻实施，对于中国野生动物的保护工作无疑产生了深刻的影响，起到了一个极大的推动作用。另外，《生物多样性公约》第十五次缔约方大会（COP15暨"昆明会"）也因疫情在全球的肆虐，从2020年10月推迟到2021年5月，又推迟到了2021年10月召开。这期间的2021年2月1日，《国家重点保护野生动物名录》在做了重大修改补充后正式公布，此书也须根据新名录作出相应调整和修改。这些都是我在构思并基本成型这本书时万万没有想到的。因此，在后记中有必要专门补上这重要的一笔。

本书的编辑出版得到了中国科学院魏辅文院士的充分肯定和支持并撰写了序言，得到了国家林业和草原局自然保护地管理司、野生动植物保护司、宣传中心和中国野生动物保护协会、中国野生植物保护协会的大力支持。本书的出版得到了广东省长隆野生动植物保护基金会的支持和赞助，得到了国家林业和草原局自然保护地司王志高司长，野生动植物保护司周志华副司长、张月处长，农业农村部渔业渔政管理局张宇主任科员，中国野生动物保护协会水生野生动物保护分会主任李彦亮，北京师范大学张正旺教授，中国水产科学研究院樊恩源研究员，广东海洋大学廖宝林教授，江西省林业科学院野生动植物研究所黄晓凤所长、刘鹏、冯莹莹以及徐永春、柳峰、甘海东等专家的大力支持，在此一并给予衷心的感谢！

另外，在几十年的资料收集、拍摄过程中，得到了各省（自治区、直辖市）林业厅（局）、野生动物保护和自然保护地管理部门的大力支持，得到了诸多自然保护区、国家公园、野生动物救护中心、野生动物园的大力支持并提供了可能的帮助，在此也一并给予衷心的感谢和诚挚的谢意！没有你们对于中国野生动物保护工作的默默奉献和保护成绩，没有你们的积极支持和热情帮助，这本书是万万不可能历经那么多年的积累，并最终完成资料收集、立意构思并编辑出版的。

书中个人的认识肯定是粗浅的，也会存在一些错误和不足，望大家多提宝贵意见。对于以上帮助，本人再一次表示衷心感谢！

Epilogue

The sudden outbreak of COVID-19 in early 2020 hit the entire globe in a severity that had never been seen over the past hundred years. China, the country where the deadly virus was detected the earliest, is also one among its global counterparts that have done the best in putting the disease under effective containment within the shortest possible time. Now that the spread of the disease is basically under control, the country has recovered from the hit on a comprehensive scale, with its productive activities and schools reopened, and people gradually moving back into normal paces in their daily lives. Meanwhile, China has also been, and is still, shipping massive amounts of resources, vaccines as well as expertise gained from its disease-combating endeavor to other countries in the world that are still fighting hard against the pandemic. These are the results of the timely, powerful and well-calibrated measures that the Chinese government has taken in response to this catastrophic moment in human history. This fact demonstrates tellingly that China has lived up to its role as a responsible major country in the world.

Shortly after the outbreak of the disease, the Standing Committee of the National People's Congress responded promptly and released on February 24, 2020 a historic ordinance — *The Decision to Place a Comprehensive Ban on All Illegal Transactions Related with Wildlife and to Wipe Out Undesirable Habits in the Abusive Consumption of Wildlife-derived Food so as to Safeguard the Health and Safety of the People* (hereinafter referred to as the *Decision*). The *Decision*, which targets in particular at the major problems related with the irrational consumption of food derived from wildlife, is meant to place, before the formal amendment of laws governing issues in this regard, a strict ban on consuming wildlife as food and crack down on all unwarranted transactions concerning wildlife, hence laying down a solid legal foundation for the urgent administrative actions taken for the purpose of safeguarding public health, eco-security as well as the health and wellbeing of the public. Indisputably, the release and implementation of the *Decision* is of great significance to wildlife conservation in the country and has pushed our efforts in this field to a new height. Moreover, the pandemic that rampaged throughout the world also put the UN CBD COP15 (also known as the Kunming Conference) off from October 2020 to May 2021, which is now rescheduled to convene in the upcoming October. Following the formal release of the newly amended *List of Key Protected Wild Animals of National Significance*, on February 1, 2021, in which major changes and additions are made to the previous list, corresponding information in this book was also duly revised. I could hardly have imagined those when I first started on the writing of this book. For all the above-mentioned reasons, I think it is absolutely necessary for me to put in this additional message in this epilogue that I am writing now.

I am most grateful to Mr. WEI, the CAS academician, for all the heart-warming encouragements and supports that he affords me as I work on this book, and also for the Foreword I he has kindly written. Supports have also been generously provided by the Department of Natural Protected Areas Management (DNPAM), the Department of Wildlife Conservation (DWC), and the Information and Publicity Center of the National Forestry and Grassland Administration (NFGA), as well as by the China Wildlife Conservation Association (CWCA) and the China Wild Plant Conservation Association (CWPCA). In addition to the sponsorship granted by the Guangdong Chimelong Wildlife Conservation Funds, the following people have each offered their helpful hands and valuable advices: Director General WANG Zhigao from NFGA-DNPAM; Deputy Director General ZHOU Zhihua and Director ZHANG Yue from NFGA-DWC; ZHANG Yu from the Bureau of Fisheries under the Ministry of Agriculture and Rural Affairs; Director LI Yanliang from the Aquatic Wildlife Conservation Branch (AWCB) affiliated to the China Wildlife Conservation Association (CWCA); Professor ZHANG Zhengwang from Beijing Normal University; Research Fellow FAN Enyuan from the Chinese Academy of Fishery Sciences; Professor LIAO Baolin from Guangdong Ocean University; Director HUANG Xiaofeng, LIU Peng, FENG Yingying from the Wildlife Research Institute of the Forestry Academy of Jiangxi Province, as well as many other specialists such as XU Yongchun, LIU Feng and GAN Haidong, to all of whom I owe my most sincere thanks.

Over the past decades that I spent in taking photos and in doing deskwork in preparation for this book, the forestry authorities in each of the provinces (municipalities and autonomous regions) across the country, the administrative departments responsible for wildlife conservation and protected area management, and many nature reserves, national parks, wildlife rescue centers as well as wildlife zoos have all done their utmost to help, to each of which I hereby extend my most sincere gratitude. If it had not been for your hard and dedicated work in wildlife protection and conservation, if it had not been for the unreserved supports and encouragements that you offered me over all the years I spent in preparing for this book, it would hardly be possible for me to complete this tough work and to present this eco-album to the readers.

The comments, ideas and proposals that I raise in this book are very likely to be poorly-conceived of or to fall short in their academic values, and there inevitably might be some errors or insufficiently-grounded viewpoints. Kind feedbacks and invaluable comments from our informed readers will be tremendously appreciated. My most heartfelt thanks go once again to all those who have helped in the making of this book that has by no means come by easily.